The GOOD BUG

The GOOD BUG

A Celebration of Insects and What We Can Do to Protect Them

GEORGE McGAVIN

Michael O'Mara Books Limited

First published in Great Britain in 2024 by
Michael O'Mara Books Limited
9 Lion Yard
Tremadoc Road
London SW4 7NQ

A CIP catalogue record for this book is available from the British Library.

This product is made of material from well-managed, FSC®-certified forests and other controlled sources. The manufacturing processes conform to the environmental regulations of the country of origin.

ISBN: 978-1-78929-669-3 in hardback print format
ISBN: 978-1-78929-671-6 in ebook format

1 2 3 4 5 6 7 8 9 10

Cover illustrations by Carim Nahaboo
Cover design by Ana Bjezancevic
Illustrations by Karen King
Typeset by Design 23

Printed and bound by CPI Group (UK) Ltd, Croydon, CR0 4YY

www.mombooks.com

Contents

Introduction

This book is about insects (known colloquially as bugs), the most successful, abundant and enduring multicellular animals that have ever lived on Earth. Together with all other species that possess an exoskeleton and jointed legs, such as arachnids (spiders, etc.), myriapods (millipedes and centipedes) and crustaceans, insects belong to a very large animal group called the Arthropoda, which accounts for more than half of all known species and around three-quarters of all animals. The biomass of all arthropods is more than 40 per cent of all animal biomass and represents 1 billion tonnes (1.1 billion US tons) of carbon. The vast majority of arthropods are insects, so we need to know about these familiar six-legged creatures in order to understand how the natural world works. After all, insects were the first animals to colonize the land and the first to take to the air. By the time our ancestors were crawling out of brackish pools on stubby, fin-like limbs to survey the land, insects had already been there for 40 or 50 million years. So, why do some of us dislike or even fear insects? In adults they can and often do provoke anything from mild disdain to all-out phobia. Yet most young schoolchildren find bugs absolutely fascinating – and some, like me, will go on to study them for the rest of their lives.

Killing insects with poisons of various kinds has long been the hallmark of our hostility towards them. Since the beginnings of agriculture, it seems that humans have been engaged in a never-ending battle against the bugs that these days involves the use of huge quantities of pesticides. It is claimed that feeding our growing population will require the use of even more chemicals, but this ignores the fact that we already waste one-third of everything we grow.

There is widespread ignorance about insects and their considerable role in how global ecosystems work. I once saw a pest-control van with a very realistic image of a honey bee painted on the side. The driver told me with utter conviction that it was a wasp and therefore fair game. As we shall see later in this book, there is no need whatsoever to go around slaughtering wasps anyway. The large black-and-yellow species that make paper nests in sheds, attics and hedgerows are generally disliked. Towards the end of summer, when wasps are no longer feeding larvae back in the nest, they seek out sugary foods to sustain themselves until the first frosts kill them off. Consequently, they can make a nuisance of themselves whenever there's food around. But with a little more understanding and a little less hysteria, we can get along with them just fine. For example, I usually take a little spoonful of jam with me on a picnic, place it some distance away from where we're sitting and let the wasps find it, distracting them from the food we're enjoying.

Bloodthirsty insects such as mosquitoes, midges and horse flies are extremely unpopular as their bites can be painful and itchy – and they carry pathogens and parasites. Yet even the tiny Scottish midge, a notorious blood feeder with a fearsome reputation, has its good side. Have you ever wondered why large parts of Scotland remain wild and beautiful? Imagine how many more golf courses, holiday homes and the like there would be if the midge were ever vanquished. To my mind, midges are a price worth paying.

I am sometimes asked what the most dangerous animal on Earth is and I try to resist saying human beings because I know that's not what people mean. The answer is, of course, the mosquito, specifically the species that act as vectors for the micro-organisms that cause malaria, yellow fever, dengue and other diseases. Hundreds of millions of people suffer from the effects of these bites, and malaria alone is responsible for nearly 1 million deaths a year – mostly of African children under the age of five. In the long history of human warfare, it is very likely that more combatants died of insect-borne diseases and infection than of injuries sustained on the battlefield. That changed with the advent of effective pesticides to control malaria and typhus, but it is easy

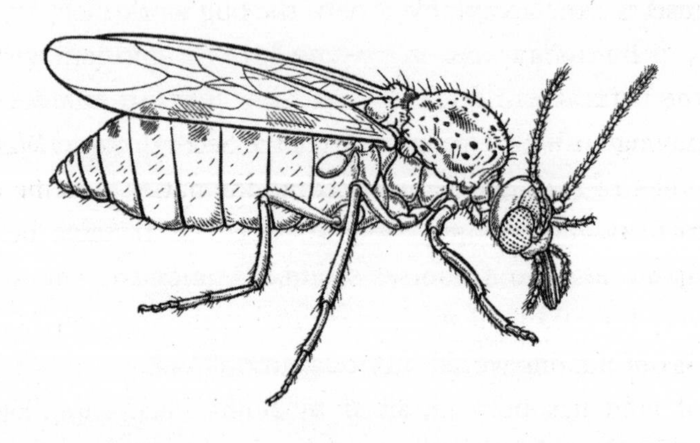

Scottish midges breed in boggy areas in northern regions from Europe as far east as China.

to understand why some insects get a bad press.

Inevitably, some insects have wriggled their way into our good books. In folklore, ladybirds are associated with love, luck and fertility; they are widely thought of as being the gardener's friend too because they eat prodigious numbers of aphids. Butterflies flit about looking rather pretty and not doing anyone any harm. Bees are the pre-eminent pollinators of flowering plants, without which there would be a great loss of human food – perhaps as much as one-third. Honey bees are also very useful to us as a source of honey and wax. These creatures are the poster girls (and boys) for insects – the acceptable face of the bug world.

But today we seem to have lost a connection with the natural world and insects have consequently been having a tough time of it. After the Second World War, when rationing was over, feeding the nation became a priority. Agriculture shifted up several gears and took up more and more land. Our priorities seemed clear and very little thought was given to what the consequences of our actions would be. Today, more than 75 per cent of land has been modified by human activities. The arrival of global super-crops such as soya bean and palm oil has had a particularly devastating impact on biodiversity worldwide.

Sadly, insects are much less abundant than they once were. The reason for this decline is the steady loss of suitable habitat and the huge quantities of toxic chemicals used in intensive agriculture.

There are people who just dislike insects, but there are many more who are simply not that interested in them. I once gave a lecture during which I showed images of some pretty spectacular tropical insects. One mimicked the moss it rested on so perfectly that it was hard to see where the moss stopped and the insect began. Another was decked out in colours so vivid it was difficult to believe it was real. At the end of my talk, an audience member asked me why he should care about a species he would never see. I told him about phytoplankton. He may never have seen phytoplankton himself, but we know that they make most of the planet's oxygen so they're something we should all care about. In discussions about insects and nature in general, it's easy to slip into very anthropocentric arguments about utility. What use are insects to us? Why should we care? Many people still think the world and all its biodiversity exists purely for our benefit. The old-fashioned notion of humankind's dominion over nature is deeply engrained, and changing this mindset will take a lot of work.

You've probably picked up this book because you generally like insects, or at least find them interesting. But, of course, the people I really want and need to read it are those who recoil in horror if nature gets too close – the ones who spray and kill any bugs they find in their homes and gardens and would rather have artificiality and sterility than the reality of our natural world any day of the week. If they don't read this book, I will be merely preaching to the converted. So, it falls to you to spread the word among your friends, family and colleagues. You need to speak up for insects because we need them far more than they need us. Actually, except for a couple of species of louse, they don't need us at all.

The Earth is 4.5 billion years old and it is only during the last half a billion years that life has progressed past a single cell. Arthropods – the majority 'stakeholders' of our planet – have dominated life in the oceans and on land for much of this time. Humans indistinguishable from us appeared a little under 200,000 years ago and the occurrence of an unusually warm interglacial period just 10,000 years ago allowed us to proliferate. We have spread across the planet and biodiversity (a term coined only a few decades ago) has taken a battering. The loss of species and natural habitat continues. The destruction of our natural capital, a direct result of our growing numbers, is very bad news indeed and things will get worse until we realize that these resouces are not infinite. Our success as a species has affected planetary ecosystems and has brought about a major extinction event that may yet rival some of the other mass extinctions that have shaped life on Earth.

Like it or not, most species on Earth have six legs and are crucial to our survival. There's never been a more pivotal time to understand their brilliance and appreciate their worth. I do hope this book can play a part in changing attitudes.

CHAPTER 1

The Nature of Insects

'Even the smallest insect, with its intricate structure, is far more complex than either an atom or a star.'

MARTIN REES

✦ The bug blueprint ✦

I'm lying under the spreading canopy of an English oak tree. The acorn from which it grew fell to the ground sometime during the reign of Henry VIII. It is late summer and the delicate, pale-green leaves that unfurled in early May are no longer pristine. They are tougher, but they are torn and tattered and their surfaces bear strange growths and pale patches. The ravages of insects can be clearly seen and as I look up at the dense mass of foliage, my mind wanders back to a time when the Earth looked very different. A time when there was little or no life on land and from the warm, shallow seas emerged small, six-legged creatures that would become the most diverse and enduring life form ever seen.

These small animals made their appearance on land towards the end of the geological period known as the Silurian and the beginning of the Devonian period around 420 million years ago. The establishment of terrestrial life at this time was a major revolution. The early insects that were part of it were scavengers and would have looked a bit like modern-day silverfish and

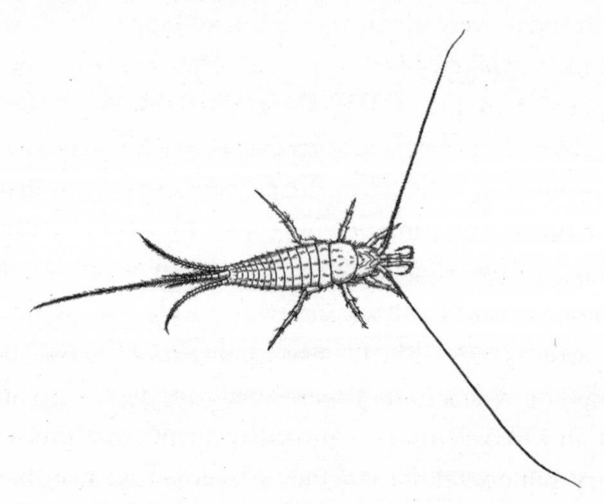

Bristletails are primitive, wingless insects.

bristletails. But they had an adaptable and versatile blueprint and in time would conquer the world.

Recent studies have shown that insects are very much more closely related to crustaceans than they are to the other land-living arthropods such as arachnids (spiders and allies) and myriapods (millipedes and centipedes) – any apparent similarities evolved as they adapted to living on land. The gradual move to land via coastal habitats and saline pools became possible partly because the spread of vegetation made the land increasingly habitable. Mats of micro-organisms, algae and small clumps of early plants unlike anything on Earth today were already growing and the early insects would have taken advantage of these new opportunities. The seas were also full of things that would eat them; the absence of terrestrial predators would have made the transition even more likely. Suitable fossils from these times are rather thin on the ground but we know enough to say that insects were a central part of the development of the Earth's early terrestrial ecosystems.

Insects were already pre-adapted to colonize the land. For a start, they were arthropods, so already had an exoskeleton that protected them from physical injury. But now that tough outer layer would also shield them from dangerous levels of ultraviolet radiation from the sun; and the very real risk of drying out could be avoided by adding a layer of waterproof wax.

Another great advantage was that insects had evolved a more compact form. The segments of the body were now rationalized and amalgamated into three major sections – the head, thorax and abdomen. This was a much more efficient scheme and allowed each group of segments to become specialized. The head carried the all-important sensory organs, the

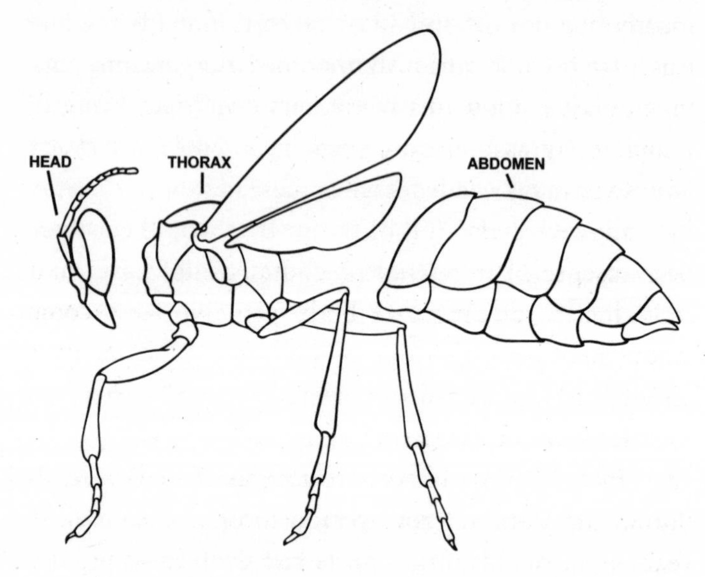

All insects have three main body parts: the head, thorax and abdomen.

antennae, the eyes and the mouthparts, which became wonderfully adapted to deal with a wide range of foods: the sharply toothed jaws of tiger beetles are used for immobilizing and dismembering prey; aphids plug their slender, elongate mouthparts into plants to suck up sap; house flies use a sponge-like structure to lap up liquids; while butterflies and moths use a flexible proboscis to drink nectar.

The middle body section, the thorax, became the locomotive powerhouse, bearing three pairs of jointed legs that became variously modified for catching and manipulating food (mantises), running (tiger beetles), jumping (grasshoppers), digging (mole crickets) and swimming (water beetles) and, later, the wings that allowed them to take to the air. The final major body section, the abdomen, contained the guts and reproductive organs. This body plan would become wildly successful, giving insects a huge advantage over all other terrestrial arthropods.

Terrestrial ecosystems gradually developed into the vast tracts of wetland forests that dominated the land during the Carboniferous period (356 to 299 million years ago). By this time, insects had evolved wings and many species had grown to a large size, perhaps because of the higher levels of oxygen in the atmosphere. The evolution of wings was a very significant advance as it allowed insects to spread far and wide in search of food and mates, and to avoid enemies and unsuitable environmental conditions. Incidentally, the next animals to take to the skies were the pterosaurs, 175 million years later.

During the Carboniferous period, insects evolved other characteristics that saw them become not only pioneers but the ultimate survivors. First, they evolved a wing-folding mechanism. Being able to fold their wings back along the body and out of harm's way meant they could take greater advantage of the huge range of terrestrial microhabitats that were now available.

They also changed the way they developed from egg to adult. The earliest insects did not undergo any metamorphosis: the young stages just looked like small adults and continued to moult even after they reached sexual maturity. (This most basic form of development is still seen in silverfish and bristletails today.) But the evolution of wings also needed some mechanism for their development. In some, metamorphosis is incomplete and the immatures, called nymphs, grow and moult several times, with the wings appearing as small pads on the outside of the thorax. During the last moult, just before the insect reaches adult stage, the wings are expanded.

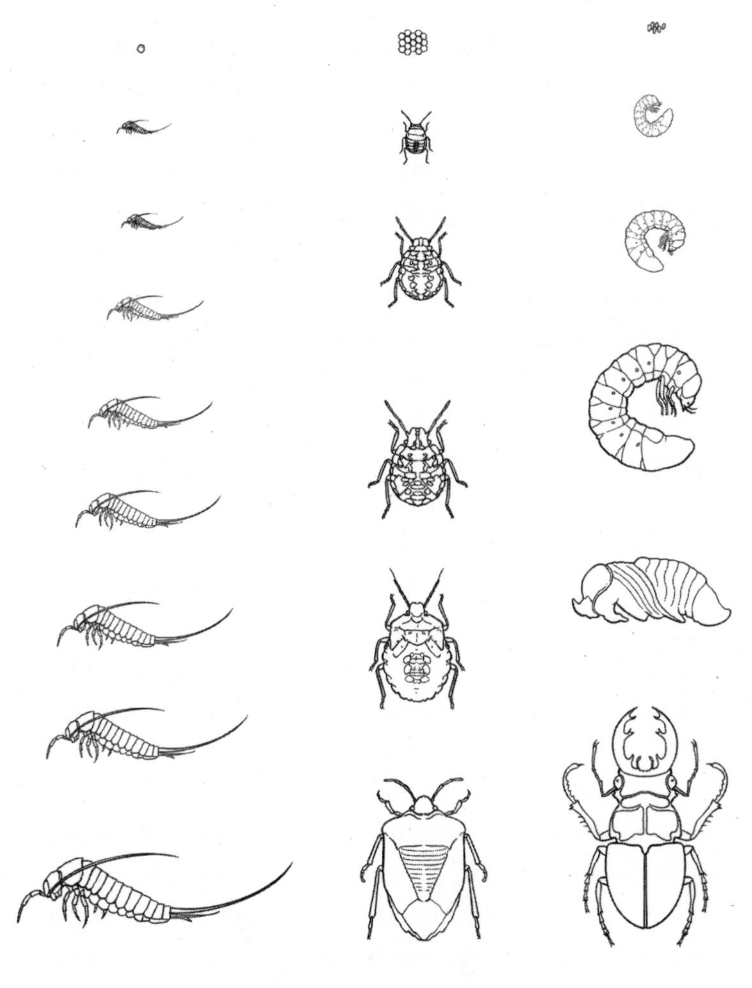

(Left to right) Development of a bristletail, a shield bug and a beetle.

This kind of development is seen today in insects such as grasshoppers, cockroaches and hemipterans.

The most advanced form of development, as seen in the biggest extant insect orders, such as beetles and flies, is known as complete metamorphosis. Here, the immature stages (larvae) are very different from the adults they will become, and there is an intermediate pupal stage inside which a certain amount of tissue reorganization takes place. The evolution of a pupal stage, which occurred around 350 million years ago, had several advantages. It allowed insects to cope with fluctuating seasons and climatic instability, and meant that the immatures and adults could exploit different resources. In effect, the larvae became eating machines while the adults became sex machines.

✦ How many species? ✦

Before we delve any deeper into what makes an insect an insect, it's helpful to know how many species we are likely to be dealing with. You would have thought that hundreds of years of inquiry and generations of biologists might have given us some answer to this basic question. The astonishing truth is that we still have only a vague idea. It was already well understood that most of the species on Earth would be small creatures because small species with shorter lifecycles evolve at faster rates than long-lived species. Except for the occasional new discovery, it is unlikely that many more species of birds, fish and mammals will come to light. But when it comes to insects we move into the realm of the unknown – and perhaps the unknowable.

When I read zoology at Edinburgh University in the early 1970s, it seemed to me that taxonomy (the science of naming, describing and classifying species) was not only important but also something we could get to grips with, given a bit of application. But it seems that students are increasingly less interested in it, and this lack of engagement is becoming a serious issue. It might at first glance seem rather dull and dusty stuff, but the naming or ordering of species is the cornerstone of biology. Without a system of classification all you are left with is an unconnected jumble of observations; and you cannot compare your results with anyone else's if you can't agree what species you are working on in the first place. Mathematics may be regarded by some as the basic science, but taxonomy certainly isn't far behind. Being able to tell one thing from another is a basic function of the human brain and this ability allowed our ancestors to make sense of and survive in a very complex world. Early humans might have counted sheep or goats, but they needed to know the difference between the two things to do so.

The first serious attempts to quantify how many living species there were took place in the 1980s, the same decade that the word biodiversity was coined. The most species-rich habitats on Earth were known to be the tropical forests found in the equatorial zone, making them the obvious place to start. The trouble was that rainforests are difficult places to work and most of the species to be found there were high up in the canopy. Getting into the canopy was dangerous and difficult. Researchers climbed up using ropes, built walkways through the canopy or erected giant cranes and even employed hot air balloons with nets slung beneath to access the biological riches that were otherwise out of reach. These techniques had their limitations, not least of which was the great cost and the relatively small area of forest canopy that could be covered each time.

Insecticidal knock-down techniques were all in all quicker, cheaper and yielded large amounts of material. A mist of fast-acting insecticide, usually a pyrethroid (an organic nerve poison derived from the flowers of Chrysanthemums) was blown into the canopy and cloth trays or funnels were set up on the forest floor to collect the fall-out. The results of these studies were quite eye-opening and pointed to there being many millions of unknown insect species. One estimate was 30 million and for a while there seemed to be a competition between scientists to see who could publish the highest figure. Things eventually settled down and a consensus began to emerge. It is now thought that there are somewhere between 8 and 10 million species on Earth and the vast majority of these are insects. The reality is that we will never know for sure, as the rate at which species are being lost – mainly through habitat destruction – means that some species

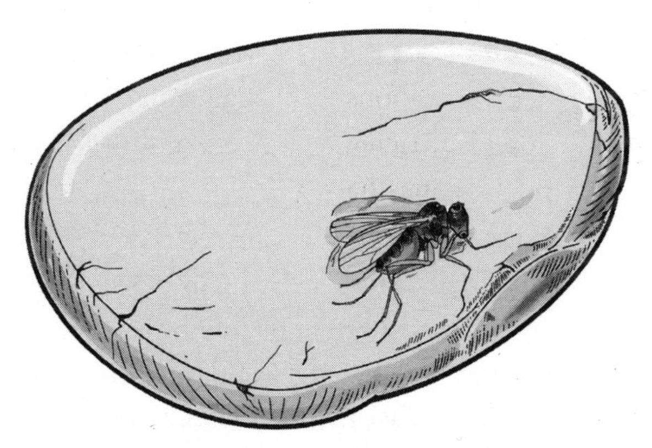

A scuttle fly, nearly 100 million years old, perfectly preserved in Burmese amber.

will appear and disappear without us ever knowing they existed. Nevertheless, insects are an astoundingly diverse group because of the speed at which species bring about new species (their speciation rate) and since many insects show relatively low rates of extinction. Most insect species may survive unchanged for 2 million years or so; but a few species found in Baltic amber that is 35 million years old are identical to species we can find today. In the long term, though, the unfortunate reality is that extinction is (and always has been) the norm on our rocky little planet. Only 1 per cent of all species that ever lived on Earth are alive today.

◆ Insects inside and out ◆

For many people, swatting a fly is the only chance they'll get to examine the insides of an insect. I must confess to having swatted a few myself. Perhaps you've taken a closer look afterwards – I'm sure significant biological careers have been launched by less. American zoologist Vincent Dethier's groundbreaking studies on fly behaviour and physiology began by his chance observation of a blow fly laying her eggs on a liverwurst sandwich someone had left on a windowsill. At first

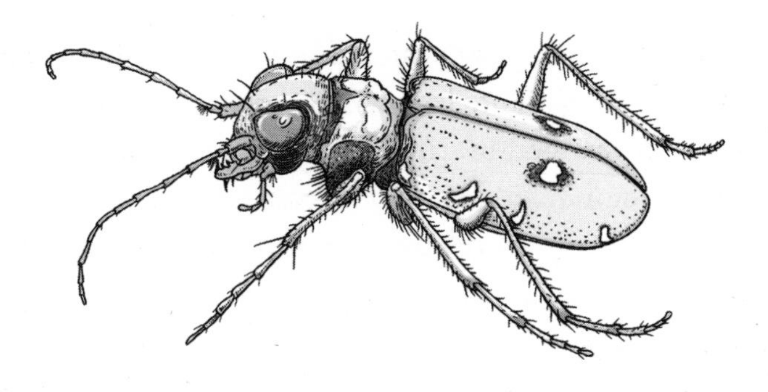

Tiger beetles are among the fastest running insects in the world.

glance, a fly's corpse might not look very interesting; but like most animals, including us, flies have a functioning nervous system – well, they do when they're alive. A network of super-sensitive air current-detecting hairs and eyes can process visual signals far faster than we do, although speed can be a problem sometimes. This is true of other insects too: tiger beetles, the fastest thing on six legs, run so quickly that their vision is blurred (a similar experience would be sitting in a moving train and watching the trackside vegetation). To deal with this, tiger beetles run in short bursts, stopping every so often to check exactly where they are.

But back to flies. They have a digestive system – a long tube running from the mouth to the anus with different sections for absorbing nutrients and getting rid of waste. There is also a circulatory system with a tube-like heart lying along the dorsal surface. The insides of insects are bathed in a fluid called haemolymph. This all-purpose liquid transport system, analogous to the blood of vertebrates, is kept in constant circulation by the heart, which pumps it from the back of the insect to the front. If you were looking at a female fly corpse, you might see a few elongate eggs that it was just about to lay. The trouble is that everything inside a fly is a sort of creamy white colour, so it can be difficult to make out what all the various bits are without the help of a good microscope.

Only the outside covering (cuticle) of the fly is recognizable. In some insects, such as the appropriately named iron-clad beetles, the cuticle can be so tough that entomologists find it very hard to push a pin through it, while in others it is extremely thin and delicate. The differences are due to the size of the insect concerned, and how it lives. Burrowing and ground-dwelling species will need to be tougher than those that spend their adult lives on the wing. As the cuticle has an obvious protective function, it might be tempting to regard it as merely a suit of armour, but it is much more than that. Its primary function is to act as an exoskeleton. Insects can move around only because their muscles are directly attached to the inside of the cuticle. The legs, for instance, are like a series of tubes joined by soft areas of cuticle that act as hinges. Muscles originating in one leg segment pass through a joint to be anchored in the next segment, and when the muscles contract the rest of the leg moves one way or another. The same goes for all the body segments.

Cuticle is a remarkable composite material made up of molecules of a polysaccharide biopolymer called chitin embedded in a protein matrix. In areas requiring extra toughness, such as the tips of jaws and claws, the outer layer of the cuticle can be chemically hardened. The cuticle is also capable of some degree of self-repair and can protect against pathogens such as bacteria and fungi. It positively bristles with all manner of sense organs, too, relaying temperature, pressures, internal strains and the slightest movement of air.

If you have ever been fortunate enough to have watched an insect moulting, then you have witnessed a process that has been repeated many trillions upon trillions of times since arthropod animals first appeared hundreds of millions of years ago. It is such a common occurrence, yet it truly is one of the wonders of the natural world. The disadvantage of having an exoskeleton is that it will not expand, so from time to time it needs to be replaced to allow for growth. Some insects moult as few as three or four times during their life, while others may do so fifty or more times. Cuticle is produced by a single layer of cells called the epidermis. When moulting occurs, the epidermis draws back from the inner surface of the old cuticle. The epidermal cells start dividing and the new epidermis, having a larger area, appears wrinkled. A moulting fluid, a mixture of enzymes, is secreted into the space beneath the old cuticle, which begins to be digested and reabsorbed. The epidermal cells can now secrete a new cuticle and the process of shedding the old one begins. Moulting can be a dangerous time for any insect because the cuticle's protective function is seriously compromised until the new cuticle is expanded and hardened.

• Small really is beautiful •

One thing about insects that is immediately obvious is that they are generally not large creatures. The smallest insects are species of parasitic wasp, being a little more than 0.2 millimetres from head to tail. This is no more than the thickness of the cuticle of large insect species and about half the size of some single-celled organisms.

The heaviest insects around are African goliath beetles, weighing between 60 and 100 grams (2–3.5 ounces), which makes them heavier than many birds. The wingspan of a few butterflies and moths is impressive (up to 28 centimetres/11 inches) but their bodies are not very massive. The longest insect species is a stick insect from China. A female *Phryganistria chinensis*, to give it its scientific name, held the record for a while with a body length of 62.4 centimetres (24.6 inches), but one of her offspring exceeded even this with a body length of 64 centimetres (25 inches). However, such monsters are rather rare and the vast majority of insects are typically around 3–5 millimetres (0.1–0.19 inches) in length.

What is rather remarkable is that the smallest insect is pretty similar to the largest when you look at them in detail. The world's tiniest beetle could sit comfortably on the claw tip of the front leg of the world's largest beetle and yet they both have the same body segments: three pairs of jointed legs, wings protected by wing cases and virtually identical internal organ systems.

Size matters and being small has several distinct advantages. Insects need less food to develop than larger animals do. They can breed much faster than larger animals and they can take advantage of a huge range of microhabitats. A single oak tree may support a handful of birds and other vertebrates while providing food and shelter to many thousands of insect species. Dozens of species of gall wasp alone attack the roots, buds, leaves, flowers and acorns. The leaf-mining larvae of flies and moths neatly graze inside the leaves. Flat bugs scuttle about under flakes of bark, feeding on fungal threads, while very many more species of bugs suck the sap or systematically empty the contents of individual cells. Longhorn, bark and ambrosia beetle larvae devour the very fabric of the tree. These herbivorous insects may in turn fall prey to various carnivores and they all may act as hosts to one or more species of parasitic wasps or flies. Nests, fungi, lichens, mosses and the rain-filled hollows left when branches have fallen all support rich and varied communities. When the tree ages and finally dies, its decaying structure will remain alive with insects for months or years to come. The smaller you are, the more fine-grained your environment is, meaning you can just pack more insects into any given volume of habitat. Insects have simply evolved to occupy many more ecological niches than large organisms ever could.

Another important aspect of size is the relative ability to avoid high temperatures. By simply walking to the underside of a leaf or the other side of a stem, an insect can move from an uncomfortable 35°C (95°F) to a much more pleasant 20°C (68°F). To understand insects properly you need to understand how the world affects them. Animals cannot change physical laws such as the acceleration due to gravity, the properties of water and the laws of thermodynamics. A bison and a beetle lead very different lives.

In trying to explain the jumping and weightlifting capabilities of insects such as fleas, ants and earwigs, we're often told that if a flea was 50 centimetres (6 inches) long, it could leap over a tall building. Another favourite is describing what feats ants could achieve if they were the size of a human being. An average ant can carry up to fifty times its own bodyweight, but if you scaled it up, it certainly wouldn't be very impressive at all. The trouble is that giant ants simply would not work in the same way – neither would giant beetles nor any other giant insects.

This all hinges on what happens as an animal gets bigger – specifically, what the change in the ratio of surface area to volume means. Imagine an animal with a length of 1 centimetre (0.4 inches). To make it easy, let's make the animal a cube. The area of one face will be 1 x 1 = 1 centimetre^2 (0.4 inches^2), so the total surface area will be 6 centimetres^2 (0.9 inches^2). The volume of the animal will be 1 x 1 x 1 = 1 centimetre^3 (0.4

inches^3). A cubic animal five times longer would have a total surface area of 150 centimetres^2 (23.3 inches^2) and a volume of 125 centimetres^3 (19.3 inches^3). A cubic animal *ten* times longer than the first would have a total surface area of 600 centimetres^2 (93.3 inches^2), but its volume would be a whopping 1,000 centimetres^3 (61 inches^3). In other words, small animals have a relatively large surface area compared to their volume whereas large animals have a large volume compared to their relatively small surface area. The volume of tissue inside an animal roughly equates to its mass – basically, the bigger they come, the harder they fall. I like to show this effect in lectures by throwing garden peas, grapes and melons high into the air. The peas and grapes are unharmed; not so the melons. Understand this fundamental principle and you will know why insects can walk on water and elephants don't jump.

Nevertheless, the jumping ability of fleas is impressive. On average, a hungry cat flea can make a long jump of between 20 and 45 centimetres (8–18 inches). As the flea is only around 3 millimetres (0.1 inches) long, their jumps are around 150 times their body length. The jump of an impala or a kangaroo is, of course, much longer but the distance covered is only a few times their body length. So, how do fleas do it? Fleas can store energy before they jump. They can then release the energy very rapidly, producing high mechanical power over a very short time. This allows them to accelerate their small bodies in under a thousandth of a second to a speed of over 1 metre per second. Grasshoppers have a similar energy-storage mechanism in their hind legs, allowing them to store energy in elastic elements and release it in a very short time – it's much like how a crossbow works.

✦ Fast breeders ✦

If there's one thing insects are very good at it's multiplying. When asked what the point of a fly is, for example, I usually reply that it's simply to make more flies. Three of the ten biblical plagues of Egypt were lice, flies and locusts, so humans have worried about the rapid breeding of some insects for quite a while. Fruit flies famously can complete a generation in less than two weeks. At the other end of the spectrum, some periodical cicadas stay underground for thirteen to seventeen years as immatures, feeding on a very poor diet of root sap. Their spectacular mass emergences have evolved to swamp all the predators in the area to

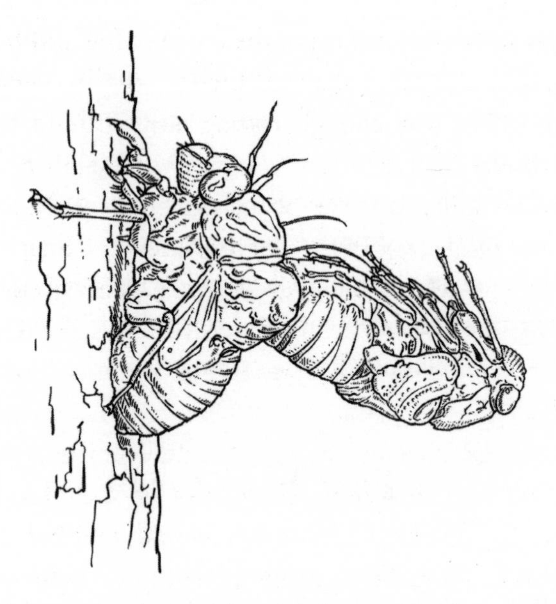

After a long time underground, a cicada nymph undergoes its last moult to become an adult.

the point that they are sick of eating cicadas. So much so that many of the cicadas crawling into the daylight to become adults and find mates (the only thing on their to-do list) will now survive.

But what would happen if insects were able to breed completely unchecked? There are many examples that demonstrate this. For instance, the descendants of a single pair of house flies could, in the space of a year, cover the surface of the Earth to a depth of several metres, assuming they all bred at the maximum rate and there was enough rotting stuff around for them to breed in. For a few years, I worked on a research project looking at two species of bean weevil and their wasp parasites. These small beetles lay their eggs on the outsides of black-eyed beans. The beetle larvae hatch and bore into the bean to feed. I reckon that a pair of these small beetles would produce about eighty offspring every three weeks or so. Theoretically, the number of beetles in the second generation would be 3,200 and, assuming there was enough food to go round, the third generation would number 128,000 beetles. By the eighteenth generation – which would, incidentally, take only 432 days – the beetles would number 1.4×10^{29} individuals. This number of beetles would occupy a volume equal to that of the Earth.

So why are we not knee deep in insects? Population explosions of insects can sometimes take place when conditions are just right, such as in non-natural situations like intensive agriculture. In general, though, insect numbers are controlled by lots of different factors, such as adverse weather, diseases, being eaten or parasitized by other animals, or their food supply simply running out. However, the astonishing reproductive potential of many insects is beyond doubt and they will multiply given half a chance.

Not all species of insect follow the route of laying large numbers of eggs and leaving their offspring to it. Some species opt for laying far fewer eggs and looking after them. Parental care is common in earwigs, for instance. Females typically lay their eggs in tunnels they dig in the soil, and groom them by licking and turning them to remove fungal spores. More than this, they will carry their eggs up and down through the tunnels to keep them at an even ambient temperature. Females also guard their eggs from predators. Maternal care continues for some time after the eggs hatch, with the

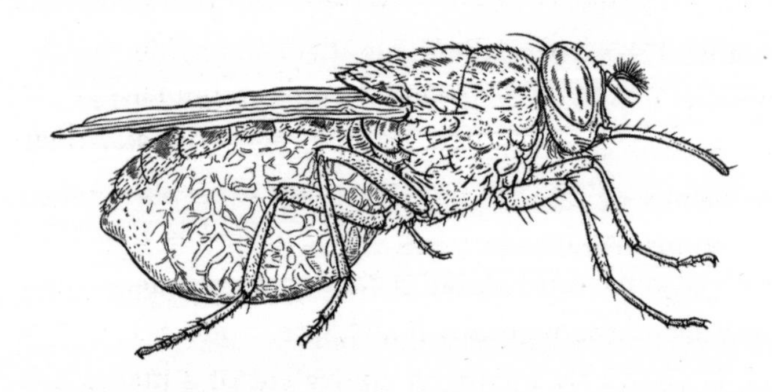

Several species of tsetse fly are vectors of the blood parasites that cause sleeping sickness in animals and humans.

mother feeding the young nymphs by bringing food into the nest or regurgitating part of her own meal. Eventually the nymphs must disperse, because as they grow older their mother stops guarding them and starts regarding them as a tasty treat.

If blow flies are a typical example of insects that lay as many eggs as they possibly can when the situation arises, the tsetse fly's reproductive strategy places it firmly at the other end of the spectrum. It has become an honorary mammal.

Tsetse flies are dull brown or grey flies found in subtropical and tropical parts of Africa. Their presence causes immense human suffering and renders large areas unfit for cattle rearing because of the single-celled blood parasites they carry and transmit when they feed. In total, some 10 million $kilometres^2$ (3.9 $miles^2$) of land are affected and around 70 million humans are at risk. Both males and females feed on blood, and their needle-like mouthparts can penetrate the toughest skin or hide. What makes tsetse flies unusual is that they produce only one egg at a time, which hatches and stays inside the mother's body. Safe inside a structure that is effectively the insect equivalent of a uterus, the larva feeds on secretions produced by special nurse or 'milk' glands. When fully grown, tsetse larvae are large and may even weigh a little more than their mothers. Larvae are then deposited on the ground, or in the host's nest or resting place, where they pupate almost immediately. After a three- or four-week pupation, the adult flies emerge from the soil and find a mate within a day or two. Females store sperm inside a special sac-like structure called a spermatheca, and need to mate only once in their lives. For insects, it doesn't really matter whether they lay thousands of eggs or only a few – it's all about survival.

• Winged wonders •

For a small animal such as an insect, flying is actually pretty easy. Think of a pollen grain; it might be carried high into the air for weeks. A small plant seed with feathery plumes can stay airborne for many days. Something bigger like an insect needs to be self-powered, but the bigger a creature gets the harder that becomes. The heaviest animal capable of self-powered flight is a bird called the Kori Bustard, which can tip the scales at 19 kilograms (42 pounds). Flying takes up a lot of energy so they don't do it very readily; and if they get any bigger they will simply become another species of flightless bird. Since all insect wings are so similar in basic structure, it is more than likely that wings appeared in their evolutionary history sometime in the Devonian period, and that by the Carboniferous there were all manner of six-legged creatures flying about.

We don't know what insect first evolved functioning wings, but it is likely that some already existing structures were modified and adapted. Solid evidence has arisen from genetic studies to suggest that insect wings evolved from appendages derived from the body wall and the bases of limbs. These structures may have been gill-like; but whatever they were, they were found in the ancestors of crustaceans and insects.

Some insect fossils show that there were more than three pairs of thoracic winglets in some ancient species. Perhaps these winglets allowed the insects to warm up quickly or aided their concealment from enemies. They might have been used for sexual display. We may never know until a lot more fossils are found, and that could take a while because fossils of very early winged insects are few and far between. Nevertheless, two pairs of winglets – one each on the middle and hind thoracic segments – became the norm. Once these proto-wings got to a certain size, they would have allowed the insects to achieve short glides or longer jumps, probably to avoid being eaten. The final step was for these structures to become fully hinged at their bases, with muscles to control their movement. Whatever the reasons for the evolution of flight, there is no doubt that it gave insects an enormous boost, allowing them to colonize just about anywhere on Earth it was possible for them to survive.

One of the factors that contributed to beetles becoming the most numerous type of insect on Earth – with 325,000 recorded species – was the development of wing cases, known as elytra. These are the beetles' front wings, which have evolved from being flight surfaces to become toughened. At rest, the wing cases enclose the hind wings, which may need to be folded once or twice to fit underneath the cases. Being able to protect their all-important hind wing means beetles can take advantage of all sorts of microhabitats that would be impossible for insects with exposed wings. Rove beetles, a very large group of elongate beetles, have very short wing cases that leave much of the abdomen exposed. This allows a high degree of body flexibility for burrowing into soil, dead wood and animal carcasses to lay eggs and raise their young. Earwigs have a similar need to be flexible and also occupy small spaces in the environment. They too have small, toughened front wings that sheath the intricately double-folded, semicircular hind wings. To help rove beetles and earwigs unfold their hind wings, some of the wing veins are elegantly spring-loaded, so packing them away needs a little bit of abdominal flexing and pushing. For some beetles living in very dry, hot environments, the wing cases have become effectively fused together along the midline of the body to reduce water loss.

Flies are the second-largest insect group, with nearly 170,000 described species. What makes them

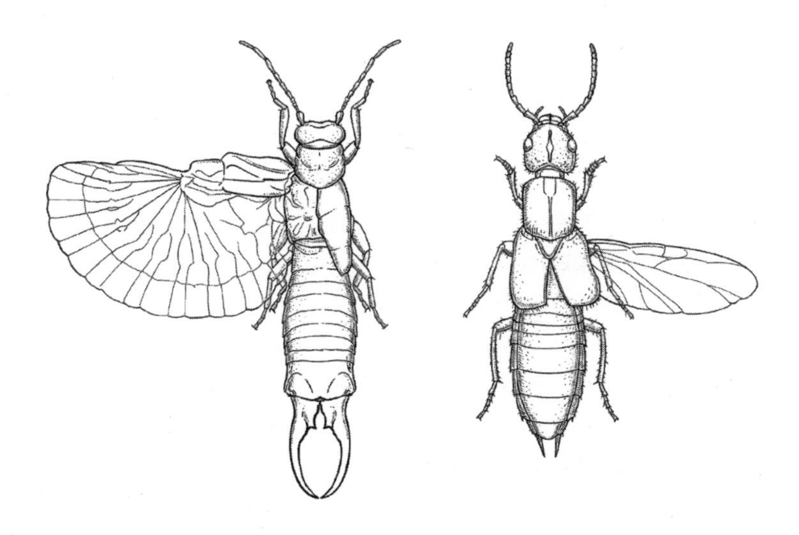

Insects such as earwigs (left) and rove beetles (right) have short wing cases under which the hind wings are folded.

different from all other insects is the loss of their hind wings, which have become reduced to a pair of drumstick-like organs called halteres. Even flies that have become secondarily wingless retain their halteres and they are quite visible in all species. Early experiments to try to determine their purpose involved simply removing the halteres from live flies to see what happened. What happened, perhaps unsurprisingly, was that the flies could not fly very well and had great difficulty manoeuvring. More elegant research revealed that these structures are miniature gyroscopes that beat at the same frequency as the wings but out of phase, providing vital positional information to the flight system.

Groups of tiny stress-receptors located in small clusters at the base of halteres respond to the minute distortions in the cuticle created during flight. These sensilla are directly connected to the flight-control system by nerves, allowing the fly to respond instantly to changes in pitch, roll, or yaw. Visual information from compound eyes and the brain is sent directly to the muscles that control the halteres, not to muscles that control the wings. In turn, signals from the halteres are relayed to the wing muscles. In this manner, the halteres, acting as flight stabilizers, can be fine-tuned by input from the eyes. The brain does not seem to have a flight-control centre; instead, this important task is undertaken peripherally by a system of super-fast reflexes.

You need only watch a hover fly feeding as it flits between flowers, even in gusty conditions, to realize how highly evolved fly flight is. Dragonflies, themselves skilled aerial acrobats, are rather clumsy in comparison to flies. The dragonfly is large and requires a relatively low wing-beat frequency to generate the lift it needs, whereas the flies, being much smaller, need a much higher wing-beat frequency to generate the lift they need. These two insects power their wings in very different ways. Dragonflies have their flight muscles attached directly to their wing bases and the nervous system can generate impulses at a high enough rate for the wings to beat at a few tens of beats per second. But flies are much smaller and their nervous system simply cannot generate nervous impulses quickly enough to operate very small wings that must beat very fast. Instead, the muscles alter the shape of the highly elastic, box-like thorax, and the wing hinges are arranged in such a way that rapid oscillations of the thorax are translated into up to hundreds of wing beats per second.

The flight systems that allow this phenomenal degree of aerial agility are complex and not yet fully understood. Hovering, backward flight, 360-degree turns and even upside-down flight and landing are nothing unusual to these insects. These skills are essential to survival, and are even brought into play during mating.

Once wings evolve, they can still be lost if having them incurs some kind of selective disadvantage – as demonstrated most dramatically among parasitic species such as lice and fleas, which spend their entire life on or very close to their hosts. Wing loss also seems to be more common in habitats such as mountains and caves, which are very stable over long periods of time. If an insect spent a lot of energy in keeping not very useful flight machinery, it would, over time, produce fewer offspring than those that diverted the energy into egg production.

✦ Insects, insects, everywhere ✦

If you are interested in some serious insect watching, the rainforests of the world are the place to go. Most insect species on Earth live in these now-dwindling tropical ecosystems. Savannahs are pretty insect rich as well and even in the short Arctic summer, insects can be plentiful.

But wherever you go in the world, there will be something interesting for you to look at. A walk along the strand line of a beach will reveal many insects, feeding and breeding in piles of washed-up seaweed and other detritus. I have recently been studying a species of small fly along the south coast of Sussex whose larvae seem to develop inside barnacles. What caught my eye initially, however, were the male flies, who dart about trying to attract females by constantly flicking a pair of highly reflective palps (sensory appendages of the mouthparts) that reflect and intensify the light even on overcast days. I don't yet know exactly how they work, but I love going down to the seaside to watch them as often as I can.

To our eyes, the sea may seem a dangerous and inhospitable place, but reasonably constant temperatures (certainly a much smaller range than found on land), coupled with fairly constant ionic concentrations, make the saltwater about as benign an environment as there is on Earth, physiologically speaking. Yet only a small fraction of 1 per cent of all insect species have returned to live in the seas they came from. There are some exceptions. In the southern hemisphere there are a few species of small, predatory bugs with the evocative common name of coral treaders. The life history of only one or two species is known, but it seems they like to hide in coral and rock crevices and emerge only to feed at low tide. A few species of caddisfly can be found in rock pools along the coast of New Zealand and south-east Australia.

Pond skaters are predatory bugs that have evolved to live on the surface of water. These slender, long-legged insects are a common sight on slow-moving streams, ponds, lakes and even cattle troughs, but another fifty or so species have lived at one time or another on the surface of the sea.

Most of these ocean striders, as they are known, are confined to mangroves and coastal regions, but at least five species are found far from land on the open oceans of the southern hemisphere. Although they live independently of land, they live *on* saltwater rather than in it. There is no physiological reason why insects could not live in saline water; but most of the features that have made insects so successful on land are not a great help to them in saltwater, so they don't. The sea is also already full of their close relatives, the crustacea, whose member species fill every available niche. Insects just don't have a chance.

The greatest numbers of marine-associated insects are found in intertidal zones, salt marshes, mangroves and mudflats, where they feed on salt-tolerant plants or are predators, or scavengers. However, these habitats are as much an extension of the terrestrial environment as they are of the marine environment.

◆ Complex relationships ◆

Insects are involved with many other species. So much so that it would be impossible to imagine any functioning ecosystem that did not have insects at its heart. They are the major herbivores on Earth, consuming more plant material than all the vegetarian vertebrates put together. They are also the largest group of carnivores.

Many insects eat each other, either directly or by laying eggs in the body of a suitable host. There are even parasites of parasites. About a quarter of all insects probably have a parasitic life history and as a result some of their lifecycles can be very complicated. Take the Black Oil Beetle, for example, which can be found across Europe and Asia. Mating for this species takes place in early spring and the male uses its curiously bent antennae to embrace the much larger female, whose swollen abdomen shows that she is full of eggs. After mating, the female lays many hundreds or thousands of eggs in a burrow she digs in loose soil. Her eggs hatch at a time to coincide with the appearance of certain solitary bees. The hatched larvae, called triungulins, have three pairs of legs, which they use to climb up the nearest flower stem to wait for the bees to arrive. When they do, the beetle larvae scramble onto the bee's back carefully to find a position where they cannot be groomed off. The solitary bee will eventually leave

A solitary bee unwittingly carries the larvae of an oil beetle back to its nest.

with a load of pollen and nectar to stock the cells in her underground nest, but she may also be carrying some oil-beetle larvae. As soon as she reaches her nest, the triungulins will crawl into the cells that contain the bee's larvae, where they will eat the provisions stored by the bee as well as her eggs and young. The fully grown beetle larvae will stay safe underground during the winter months and pupate early the following year, when they will emerge and start the whole cycle again.

When it comes to straightforward slaughter, ants excel. Ants are typically carnivorous and some species can be incredibly efficient. South American army ants or African driver ants go hunting in raiding parties, catching and killing large numbers of invertebrates as well as overpowering small vertebrates to feed their large colonies. But ants can also form symbiotic and mutualist relationships with many other species.

You may have noticed that colonies of aphids feeding on plants will attract the interest of ants. Look closely and you will see that the ants are collecting droplets of clear liquid excrement from the aphids' rear ends. The liquid is called honeydew and the ants adore it because it contains sugars. Just like every other animal, what growing insects need is nitrogen to make proteins. But plant material is not generally very nutritious, having at most 4 per cent of its dry weight as nitrogen. Plant sap is even worse, with less than 0.5 per cent weight to volume of soluble nitrogen – although it does have up to 10 per cent weight to volume of sugars. Sap-sucking bugs must consume a large volume of sap

Bullhorn acacia has a symbiotic relationship with certain ants that live inside its swollen thorns. The ants protect the tree in return for protein-rich leaf-tips and nectar.

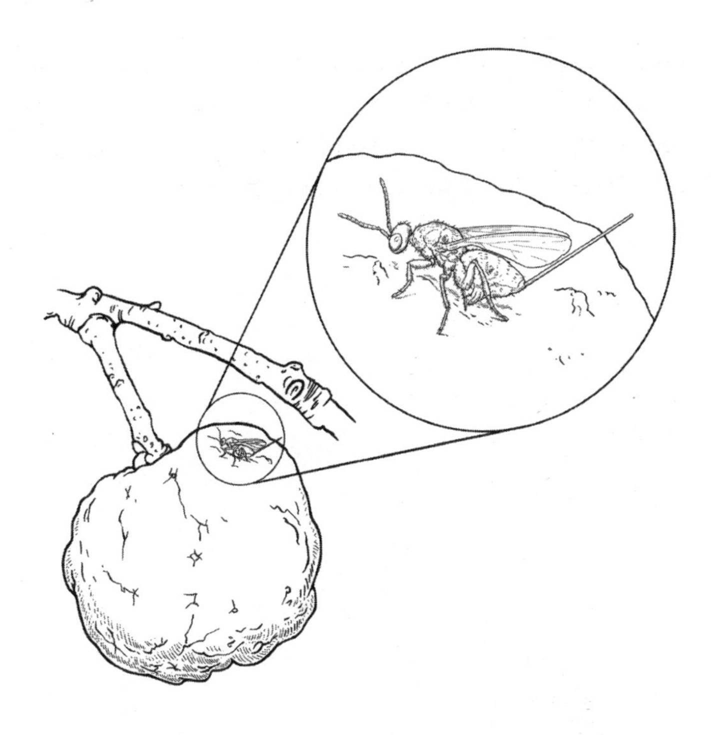

The gall wasp larvae inside this oak apple gall are not completely safe as some parasitic wasps have an ovipositor long enough to reach them.

to get the nitrogen they need to grow; and consequently must eject a large volume of sweet excrement. They even have a special modification of their gut that can divert all the water and low-molecular-weight molecules straight to the rectum – where the ants will be waiting.

Some ants have effectively become farmers, looking after the insects on whom they depend by moving them out of harm's way or onto a healthier plant where they can feed. They sometimes even build small shelters for them.

Many ants have mutualistic relationships with species of plants, which may provide them with special places to live as well as protein-rich structures and nectar for them to feed on. This is costly to the plant but in return the ants will see off hungry herbivores and may do a bit of weeding, too, so that their home is not smothered by vines or other plants. Some plants even grow large swollen structures called domatia that are full of minute tunnels and chambers in which their symbiotic ants live. The ants get a nice safe place to call home and the host plant gets the ants' waste products as a fertilizer.

In the case of gall wasps, it's not so much 'You scratch my back and I'll scratch yours', but more of an outright hijack. They induce plants to make structures that they would never normally make – galls, or swelling growths on the tissue of the plant that act as protection, nursery and larder for their larvae. Plant galls are a common sight on some plants, such as oak trees, and vary enormously in size, colour, texture and location.

In response to a female gall wasp laying her eggs, the host plant develops a gall around it. The mechanism of gall induction is not yet understood but it does seem as if the gall wasps might be genetically engineering the plant's DNA at an early stage of development. However it happens, it is always beneficial to the gall former and a net cost to the plant. Gall wasps and their galls often support a rich and varied community of interdependent organisms, many of which are parasitoid wasps that attack the gall formers and each other.

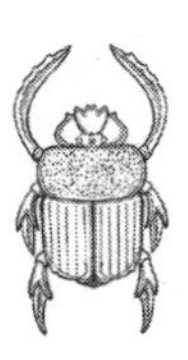

CHAPTER 2

Insects and Humans

'Nature will bear the closest inspection. She invites us to lay our eye level with her smallest leaf, and take an insect view of its plain.'

HENRY DAVID THOREAU

What is your earliest memory of an insect? Mine is of the clusters of caterpillars that I found feeding on the nasturtiums outside my maternal grandparents' kitchen window. I turned over leaf after leaf and there they were, busily munching away and all the while producing little pellets of dark droppings. I collected a few of the distinctively marked black-and-yellow caterpillars in a jam jar and added some leaves so that I could watch them more closely.

It was a little later that I found out that they were caterpillars of the Cabbage White Butterfly, and later still when I realized why these voracious devourers of all manner of *Brassica* plants were tolerated in the neatly ordered garden. My grandfather grew his own vegetables and this large patch of nasturtiums had been planted to act as a lure for the female butterflies, which preferred to lay their eggs on the nasturtiums rather than on Grandpa's kale and cabbages. It also attracted other insects he didn't want on his crops – aphids, such as blackfly and greenfly. This technique,

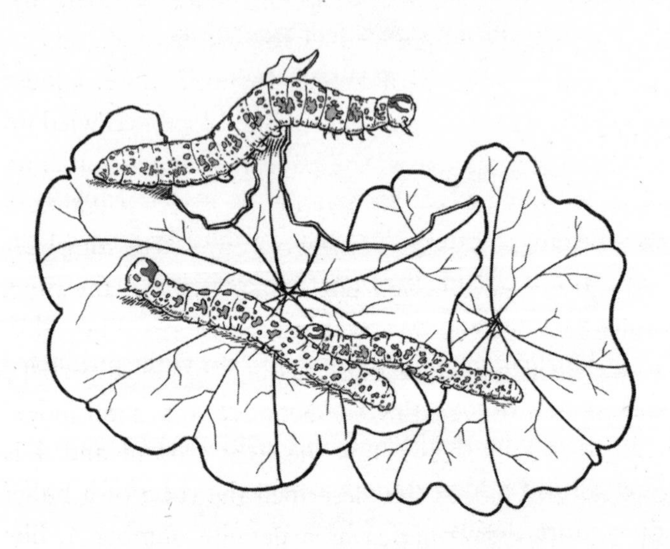

Caterpillars of the Cabbage White Butterfly feeding on nasturtium leaves.

known as the push-pull system, is a mainstay of organic farming. Plants that repel certain insect herbivores (the push) are grown alongside main crops, as well as other plants such as nasturtium that are attractive to them (the pull). So, there's really no need to douse everything with insecticides that will kill not only the insects you don't want around but also all those you *do* – the ones that would eat or parasitize the 'pest' insects.

Perhaps you had a less pleasant introduction to insects. Being stung by a wasp or a bee at a young age is something people remember for a long time and can colour their opinion of insects for ever more. I must have been seven or eight when a bumble bee crawled up my leg and stung me on the knee. I had been watching it carefully as it made its way up my sock, struggling a little as its claws caught in the wool fibres. The sting hurt and my father, although not even present at the time, told me without hesitation that I must have annoyed it.

American entomologist Justin Schmidt is perhaps best known for creating the Schmidt sting pain index, a 4-point scale where 1 is the least painful and 4 is very painful indeed. He described the sting of a bullet ant from South America as a definite number 4, like 'walking over flaming charcoal with a three-inch nail embedded in your heel'. These ants even feature in coming-of-age ceremonies among certain indigenous tribes in this part of the world. Gloves or mats woven from leaves are studded with live bullet ants and thrust onto the initiates' bare flesh. The initiates are thirteen-year-old boys and are not considered men until they have endured this torture.

Of course, some insect stings can be fatal if you have an allergic reaction and we need to be careful. But despite what you may have heard or assumed, insects such as social wasps and bees do not simply attack us for the hell of it. They sting only to protect themselves

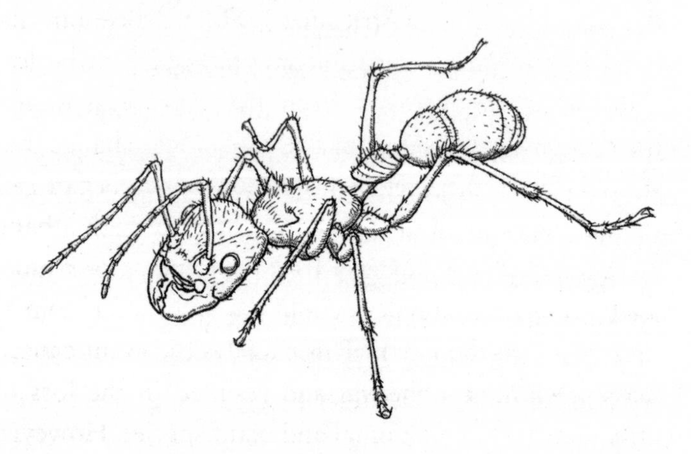

Native to parts of Central and South America, bullet ants forage for prey and nectar on forest trees.

or their brood from a perceived threat. Indeed, when bees are swarming, they are actually in the process of establishing a new colony and are very docile. In the summer, however, when the colonies are in full swing, it's not a good idea to get too close to wasps. Just like bees, they have their young to protect and will sting an 'attacker' multiple times. But they are not much bother if left alone to do what they do best, which is flying out on thousands of foraging trips to grab as many insects as they can to feed to their young.

Bees are probably regarded with much less alarm thanks to their long association with humans. Despite the menace of the Africanized Honey Bee (an ill-fated hybridization experiment), they seem to be less aggressive. Bees evolved from predatory wasps and their ancestors were part of a revolutionary change that started during the Cretaceous period, which began 145 million years ago and ended quite literally with a bang 66 million years ago when a 10–15-kilometre (6–9-mile) asteroid (some think it could have been a comet) slammed into the Earth. This cataclysmic event caused massive climatic upheaval and resulted in the loss of three-quarters of all animal and plant species. However, the really important thing about the Cretaceous period, which had been relatively warm, with vegetation stretching from pole to pole, was that it saw the flowering plants and the insects that evolved to pollinate them rise to dominance. This brought about a massive change in terrestrial ecosystems, boosting biodiversity to the point that the land was more biodiverse than the oceans for the first time in the Earth's history. Today there are around 300,000 species of flowering plants and we are almost totally dependent on them.

✦ Aliens from inner space ✦

I used to watch families taking their children around the Oxford University Museum of Natural History, where I worked for twenty-five years. In the entomology displays, I had included lots of real specimens as well as some very large and realistic models of species representative of each of the insect orders. There were also large glass tanks with live insects such as hissing cockroaches and giant stick insects. Children loved them and would tug at their parents' sleeves to 'Come and look at this!' But the faces of some of the parents showed thinly veiled disgust bordering on outright revulsion. It's clear that we are not born with a natural dislike of insects, but rather it is learned.

Perhaps our modern aversion to insects can be in part traced back to their portrayal in film and TV. There was a particular horror-film genre in the 1950s where insects – usually super-sized – were hellbent on destroying humans by whatever means they could. They scaled tall buildings, stopped locomotives and generally made a great nuisance of themselves. The hordes of giant ants in *Them!*, screened in 1954 (the year I was born), were billed as an 'endless terror and a nameless horror'. In other celluloid fantasies there were giant mantises and killer moths – even bees, the most universally beneficial of all insects, have not been safe from vilification. In many of these films, the giant creatures were immune to conventional munitions but eventually succumbed to new pesticides that were being produced at that time. In hindsight it seems quite likely that the public were being encouraged to accept the rollout of these novel chemicals, many of which were later banned because of their toxicity and the immense harm they caused in the wider environment.

✦ Moths and myths ✦

Humans have long sought to understand the world around them and stories arose to provide explanations for natural phenomena. Many of these myths involve animals of all kinds, including insects. One of my favourite creation myths is that dry land was created by a diving beetle bringing up mud from the watery depths (there must have been an awful lot of them).

In Ancient Rome it was believed that the soul left the body through the mouth and was sometimes shown in the form of a butterfly or moth. Some of these superstitions persist to this day. In some parts of the world, a large black moth entering a house is believed to be the sign of an imminent death; and the belief that the spirits of recently deceased relatives return in the form of moths or butterflies is common in Chinese folklore. In South America many believe that certain large, graceful damselflies, known colloquially as helicopter damselflies, are the spirits of recently dead humans. These slender and elongate species breed in tree-rot holes and, as adults, are able to hover near spider webs and pluck out trapped prey to eat. I suppose if you're going to believe that an insect is the spirit of your dead relative, you're going to want it to be something impressive – not some small, bloodsucking fly or a beetle that scuttles about in cow manure.

✦ Bug grub ✦

Following our appearance some 250,000 years ago, modern humans spent all but the last ten thousand years or so living as hunter-gatherers. Insects would have been a very significant part of our natural landscape and experience during that period and would certainly have formed an important part of our diet, as many still do today. My first experience of eating insects was in Papua New Guinea, where I was offered roasted beetles. They were large and crunchy on the outside and soft on the inside with a vaguely nutty flavour, but the spines on their legs stuck in my teeth (the market trader hadn't said so when I bought them but it's likely I was supposed to remove the legs first). Since then, I have gone on to try many insects and have even advocated insect eating (entomophagy) as they are an excellent source of protein for our burgeoning population. Gram for gram, termites, crickets or silk-moth caterpillars provide more calories and protein than conventional meat such as beef, chicken or salmon do, as well as significant amounts of iron, calcium and other essential micronutrients such as niacin, riboflavin and thiamine.

Insects convert plant material to insect protein much more efficiently than domesticated vertebrates and can be reared in very large numbers – so why, when over 2,000 species of insects (and some spiders) are eaten worldwide today, do Western cuisines not feature them more? Perhaps the answer has a little bit to do with culture and a lot to do with ecology. Optimal foraging theory predicts that insect eating will be worthwhile only under certain conditions. If you live in cool, temperate regions, where insects are small and not very abundant, you will use up a lot more energy collecting enough insects to dine on than you will get back from eating them.

Insect eating is slowly gaining ground in the West, but is still seen as something of a novelty to be wheeled out in the media occasionally and then forgotten about. We may need to take it a lot more seriously; but whatever happens, protein farming is going to need to get a lot more efficient in the future. My own advocacy for insect eating is waning as I now realize that what we eat is not the problem; rather, there are simply too many of us for our small planet to support. Our demands on the natural world now far exceed its capacity to meet our needs (or greed).

+ Plan bee +

The earliest representations of insects are from pictures drawn on cave walls. The oldest may be a carved or scratched sketch of what is thought to be a cave cricket. Dated to approximately 14000 BCE, it was found along with incredible clay sculptures of bison in a small cave system near the commune of Montesquieu-Avantès in the South of France in 1912.

Bees often feature in early insect artwork and the best-known examples are from Spain. One such depiction, dated around 8,000 years ago, is from caves near Valencia and shows people gathering wild honey from what appears to be a cliff face. To take risks like this as they still do in many parts of South and Southeast Asia when gathering the honeycomb of the giant or rock honey bee, *Apis dorsata*, requires a lot of nerve. The exposed combs of these species are often located in difficult to reach places and the bees themselves are notoriously aggressive in their defence of the colony and the supplies of honey they have stored. Despite this,

This rock painting discovered in the Cueva de la Araña (spider cave), near Valencia is believed to be the oldest depiction of humans gathering honey.

the nutritional rewards and pleasure of eating honey are more than enough to make it worthwhile. Wild chimpanzees routinely search for bee colonies and will use leaves or sticks to access the honey.

It must be assumed that humans have eaten wild honey for a very long time. Although there is evidence from 40,000 years ago that beeswax was used to fix arrowheads to shafts, the keeping of bees is altogether more recent and perhaps dates to 8–9,000 years ago. There are eight species of honey bee and are all native to the Old World – but one, the Western Honey Bee, has become cosmopolitan. It began to be shipped around the world in the first quarter of the seventeenth century and by the mid-nineteenth century had been imported to North America, Canada and Australia.

Honeycomb contains bee larvae and honey and today we use a wire grille to exclude the slightly larger, egg-laying queen from accessing certain parts of the hive. This ensures that the comb containing honey and the comb containing bee larvae (the brood) are kept separate, although eating both together would give you the best nutrition possible. I never tire of pointing out to children that honey is actually bee vomit, as it is nectar that has been swallowed by worker bees and regurgitated several times before it matures into the familiar substance we spread on our toast. I don't think we quite appreciate the amount of work that bees do only for the fruits of their labours to be stolen on an industrial scale. The workers from one hive may visit 1 million flowers and cover 400 $kilometres^2$ (154 $miles^2$) in a single day. A 450-gram (16-ounce) jar of honey is equivalent to about 10 million bee-to-flower forays.

Ancient hives, which did not have a queen-excluding grille, were simple wooden or clay containers, so it's very likely that bee larvae were eaten as well as the honey. Pottery containers for keeping bees dating from 5000 BCE have been found in the Middle East, and in Ancient Greece long cylindrical hives fashioned from mud or clay were used to contain bee colonies. Paintings of similar hives and the men who looked after them have been found on tomb walls of the Ancient Egyptians. Some Egyptian beekeepers still use clay pipes, building them up into walls with mud-mortar between the pots to secure them.

The Mayan civilization of Mesoamerica certainly knew about beekeeping, but kept stingless bees, which are very close relatives of honey bees. Stingless bees do in fact have a small sting but it's not in the same league as that of a honey bee. The bees would have been found in the wild and their colonies scooped out and moved into

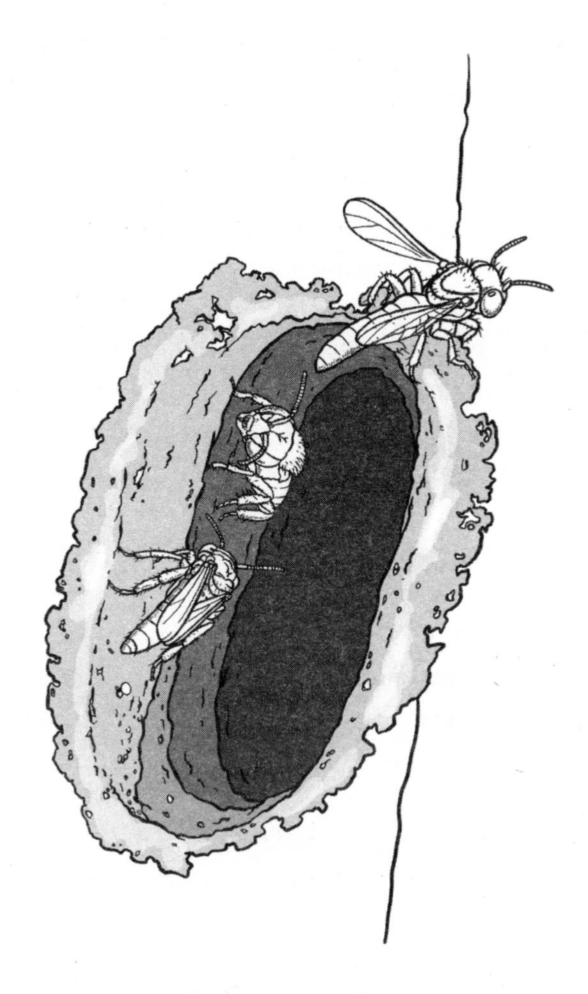

Stingless bees protect their colony by guarding a single nest entrance which is sealed at dusk.

hollowed-out logs, sealed at the ends with a wooden or stone disc. In this way they could be transported back to a village where a hole would be drilled in the top of the log to let the bees forage. Wood decays, of course, but very many of the stone sealing discs have survived and some found in Belize provide the earliest evidence of beekeeping in the New World.

The Ancient Egyptians, for whom the activities of nature ran in parallel with events in the spiritual realm, held bees in high regard. The hieroglyph for the bee was the symbol of Lower Egypt and that of kingship. The Egyptians also believed that the tears of the sun god, Ra, were transformed into bees when they touched the ground. Ra's followers were thus rewarded with honey – which was used as food, in funeral rituals and medicinally as an antimicrobial agent to treat wounds (as it still can be today) – and beeswax, which was used in the production of cosmetics and as a model-making material. The fly was a symbol of valour and prowess, and the grasshopper signified beauty and life – although the locust could refer to the size of an army.

+ Holy scarabs +

My favourite human-made object of all time is on display in the British Museum in London. It is a massive sculpture of a scarab beetle (a type of large dung beetle) carved from a single block of diorite, an igneous rock formed by the slow cooling of molten rock underground. It is 0.9 metres (3 feet) high, 1.53 metres (5 feet) long and 1.19 metres wide (4 feet), and weighs around 4.5 tonnes (5 US tons). It was made around 4 BCE and is monumental proof that scarab beetles were a very big deal in Ancient Egypt.

Khepri, a scarab god representing Ra in his morning form, signified the creation of life from nothing and, by extension, continuous rebirth. Being keen on drawing parallels between their world and the spirit world, priests must have been familiar with the lifecycle of scarabs and almost certainly made careful observations of their biology. The way large scarab beetles rolled balls of dung across the ground and buried them was an Earthly representation of the movement of the sun as it rose every day, was pushed across the sky by Khepri and disappeared below the horizon. On digging down to find the beetles' dung balls, the priests would have opened them to find new life inside and they might have witnessed the eventual emergence of a fresh, new dung beetle. It's a powerful metaphor for rebirth and everlasting life and it seems quite possible that the elaborate wrapping and entombing of pharaohs and other VIPs could have been an imitation of a pupating scarab. I do wonder, though, what they would have come up with had they known about the life cycle of parasitoid wasps.

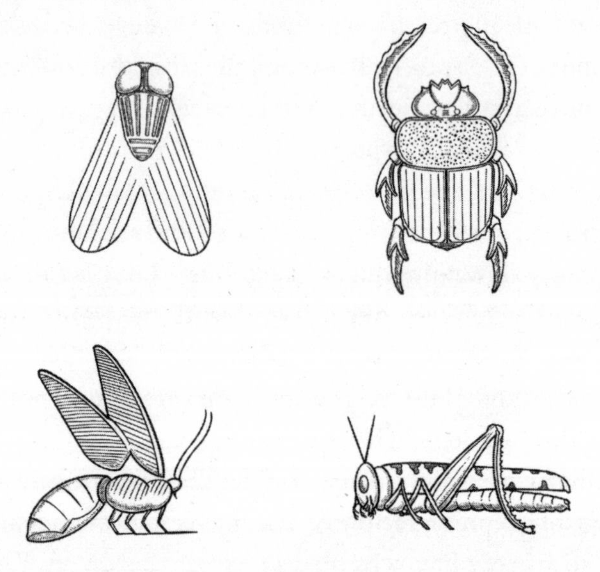

The fly, scarab beetle, bee and grasshopper were important symbols in Ancient Egypt.

Many dung beetles are not as impressively large as the sacred scarab revered in Ancient Egypt, but they're all-important to nature's recycling system. There are about 7,000 species worldwide and they dispose of prodigious quantities of animal dung. Some dung beetles live inside dung, some burrow below it and some roll balls of it away to bury elsewhere. But all of them use dung as a food source for their larvae. I once calculated that the world's vertebrates produce somewhere in the region of 100 billion kilograms (220 billion pounds) of dung every day. Whatever the exact figure, dung beetles remove and recycle an awful lot of it.

On the savannas of Africa, competition for this resource can be intense. A single elephant-dung pat can attract as many as 4,000 beetles in half an hour. A deposit weighing 1.5 kilograms (3.3 pounds) can be completely cleared in two hours by 16,000 dung beetles, themselves weighing just shy of half a kilogram (1 pound). If there were no dung beetles to do this, the dung would remain on the surface and break down very slowly, eventually causing the disappearance of the rich vegetation on which the grazing vertebrates depend. When these same vertebrates die, their carcasses will be disposed of by hordes of other insects.

Waste disposal

Blow flies are the main group of flies associated with dead and decaying animal remains as well as with dung. The adults can also be pollinators and several specialist plants attract them with flowers that smell very strongly of carrion.

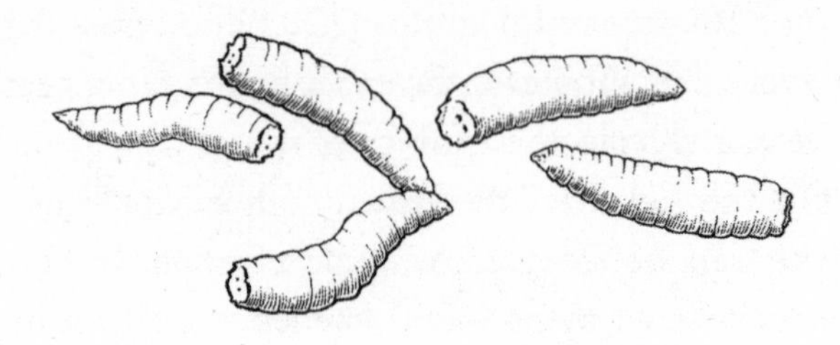

Blow fly maggots are vital in carcass disposal and nutrient recycling.

Until the seventeenth century, it was thought that maggots that appeared in dead animals came about by a process called spontaneous generation – where the dead meat somehow gave rise to living flies. It seems unbelievable to us now that this theory held sway for almost 2,000 years until 1668, when Francisco Redi, the Italian founder of experimental biology, proved that maggots hatched from the eggs of flies.

Wherever there is death, flies will follow. Flies and other insects feeding on carrion arrive in a predictable succession associated with each stage of decay. You might be put off by the nauseating smell of putrefaction but get close and you will see how perfectly adapted fly maggots are for turning dead flesh into flies. At the head end there is a pair of toughened structures called mouth hooks that rip and tear into the food. The body contains a pair of large salivary glands that pump out enzymes, reducing the food to a soup-like mush. The body is muscular, with bands of raised welts that act like the spiked shoes athletes wear and help the maggot wriggle through the sloppy food. At the rear end are two noticeable dark structures, which are the openings of the tracheal system; their location at the back end allows the maggots to feed almost continuously. Given ideal conditions, maggots can consume 60 per cent of a human corpse in a week.

Maggots can be useful in medicine and their use in cleaning up infected wounds is an ancient technique. The advent of antibiotics meant that maggot therapy, or biosurgery as it is also known, became much less popular. But now that many bacteria have become resistant to multiple antibiotics, biosurgery is having something of a renaissance. Not only can the specially bred sterile maggots eat dead tissue much faster and more neatly than a surgeon's scalpel can cut; their secretions also have an antibacterial action that can greatly reduce the numbers of potentially dangerous bacteria present in the wound.

✦ Ecosystems ✦

I often get asked what the point of a fly or a wasp is. I hate answering questions like this because they show a deep lack of understanding of the natural world. The insect under discussion has probably been around for tens of millions of years so is clearly a successful species that fills a particular ecological niche (its place within a community of species and the resources it uses). Then there's the talk about what the insect does for us . . . but of course insects don't do anything for us – that's simply shorthand for how we benefit from what they do.

Take cockroaches as an example. Just the name is likely to cause some people to wrinkle their noses. Cockroaches are often seen as rather dreadful insects that infest our homes and scurry about spreading filth and disease, but this view couldn't be further from the truth. Only a small handful of the 7,000 or so known cockroach species have become associated with humans – and we have only ourselves to blame for that relationship, thanks to our unhygienic habits and the amount of food waste we generate. Most cockroaches are essential omnivores and scavengers of a wide range of dead or decaying material, including bird guano and bat droppings, and as such play an essential part in the recycling of nitrogen.

We know that insects have been fundamental to the functioning of terrestrial ecosystems ever since complex life forms appeared on land around 430 million years ago. Modern humans' rise to global dominance over the last quarter of a million years has been nothing short of meteoric. Bear in mind that this has taken place during an unusually warm interglacial period and was helped considerably by some genetic tweaks that allowed our brains to grow three times larger than those of our ancestors. Understanding the details of our evolution might give us perspective, but what's important now is how we see our relationship with the Earth and all other species.

To put that into context, it can be useful to think about how ecosystems work. Imagine them as a layer cake in the shape of a pyramid. On the bottom layer sit the green plants, accounting for a little over 82 per cent of total biomass on Earth. They are known as the primary producers. These are the species that trap the sun's energy that makes the whole web of life work. The energy that the plants trap to live and grow becomes a food source for the animals that occupy the smaller layers above. It might come as a bit of a surprise to learn that all animals combined account for only 0.4 per cent of the total biomass on Earth and that most of those animals are insects. The second, much smaller layer above the plants is occupied by all the herbivores and is completely dominated by insects. In any terrestrial habitat you care to name, herbivorous insects are eating far more plant material than all the grazing vertebrates combined. Of course, plants have evolved many ways of defending themselves from this constant onslaught; and many of the chemicals they produce to deter insects from sucking and chewing them to death are useful to us too. These chemicals form the basis of at least a quarter of current prescription drugs – including several successful anti-cancer drugs – and other useful compounds that we use in various ways.

The next couple of layers are very much smaller and represent the carnivores – species that eat the herbivores and that eat each other. At the top of the pyramid are the apex predators – species that are not preyed upon by other species. The lion is an apex predator in Africa but it and all the other vertebrate carnivores combined are no match for ants. Collectively, ants are the biggest consumers of animal flesh on the planet and make up to 25 per cent of the total animal biomass,

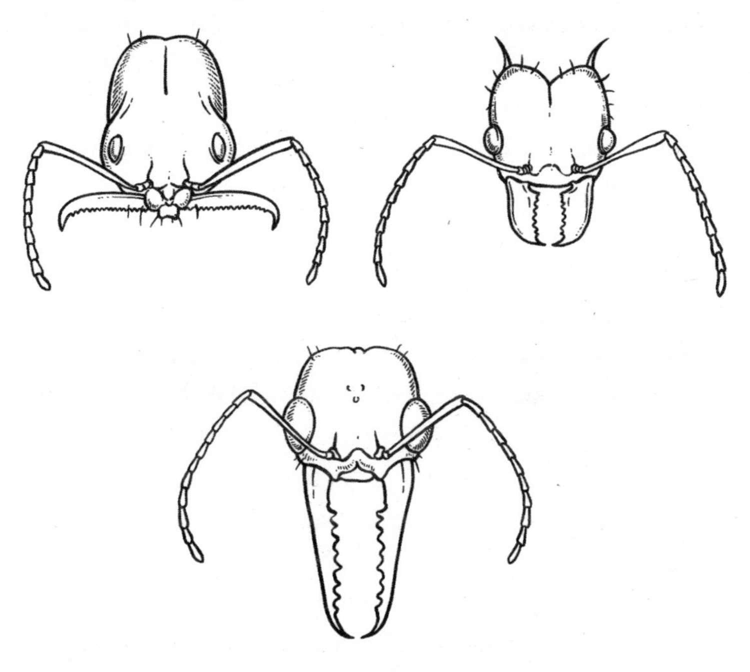

The jaws of ants are specialized for the type of food they eat.

whether you are in the tropical forests of Borneo, the grasslands of South America or your own backyard.

The reason that successive layers of the ecological cake get smaller and smaller, with increasingly less biomass, is that the amount of available energy is reduced in each successive layer. In fact, because energy is lost through metabolic processes and heat, only about 10 per cent of the energy contained in one layer is passed on to the layer above. You simply can't have large numbers of carnivores without enough prey to eat, just as you cannot have large numbers of herbivores without sufficient available plant food.

So far, so good, but this metaphor of a pyramidal layer cake is a bit simplistic because there are lots of other things going on that muddy the waters. There are parasites that live off (but don't kill) the animals they attack, there are blood feeders that just siphon off what they need and there are species called parasitoids that develop inside other insect hosts and eventually cause their death. Many species of wasps and some flies are parasitoids.

If you find a colony of aphids, you will see that some of them look a bit odd. They're immobile, seemingly stuck to the plant, and have a rounded and papery appearance. If you are really lucky, you might get the chance to see a little trapdoor being cut in the top of a moribund aphid and a small wasp emerge. Had I been lucky as a child, I might have seen the cocoons of my grandparents' cabbage white butterflies being attacked by a wasp as she laid her eggs inside them. There are even smaller wasps that are parasitoids of the first wasp – it's complicated. This sort of thing happens all the time in nature and when you put a pupa of some insect inside a little pot and rear it to adulthood, you will often be surprised at what eventually pops out.

The oak tree I mentioned at the very beginning of the book is a microcosm of ecology. From the tips of its roots to the top of the canopy, an oak tree is an ecological engine, thrumming with life. It would take many lifetimes of study to fully understand all the interrelationships between the insects, bacteria (which make up over 12 per cent of all biomass on Earth), fungi and the like that inhabit it. All the herbivorous insects munching the oak tree may end up being eaten by a number of predators; and they may act as hosts to one or more species of parasitic wasps and flies. Nests, fungi, lichens and the rain-filled hollows left when branches have fallen all support special insect communities. Biofilms of bacteria growing on the surface of leaves are grazed by fly larvae. When the oak tree ages and finally dies, its decaying structure will be alive with insects for many years until it all disappears into the soil.

+ Food of the world +

One of the most important things about insects in ecology is that they are the main food for so many other species. They really can be regarded as the food of the world.

Most birds feed their young entirely on a diet of insects. An average brood of nine Great Tit chicks may consume around 120,000 caterpillars while they are in the nest. A single swallow chick may consume many tens of thousands of bugs, flies and beetles before it fledges; and buntings, which are seed feeders as adults, rear their young on a protein-rich diet of insects. And it's not just birds. Bats rely on catching flying insects in large numbers. Many freshwater fish eat insects and a large number of terrestrial vertebrates eat insects, sometimes exclusively. Termites and ants together make up about half of the insect biomass on Earth, so it's unsurprising that several vertebrate animals have evolved to eat them exclusively. Without insects, many animals simply would not exist. Even primates (ourselves included) eat insects. The ancestors of all primates were almost certainly small, nocturnal species that lived in trees. They had large brains, forward-facing eyes and dexterous digits to catch their food – insects.

• Dressing up •

Silk is produced by the salivary glands of certain caterpillars to spin their cocoons and one of the strongest natural fibres. The best-known and most widely used silk-producing species is the silk moth or silkworm (*Bombyx mori*). The production of silk textile originated in China more than 4,500 years ago; and despite the considerable efforts of the Chinese to guard their valuable secret (silk was worth more than gold), silk moth eggs and seeds of their mulberry food plant were eventually smuggled out. Silk moth caterpillars have been bred in captivity for so long that they can no longer survive in the wild, and are kept in large, airy drawers and fed huge amounts of mulberry leaves. When fully grown, the caterpillars spin a cocoon in which to pupate and these are collected and boiled. A single cocoon may be made from many hundreds of metres of raw silk and the silk of several cocoons must be twisted together to make a single silk thread. After the silk is reeled off, the now boiled pupae are canned and sold as food. (I wouldn't recommend these – they're possibly the least appetizing insect food I have ever eaten.)

As well as being beautiful, with a lustrous sheen and a soft, luxuriant feel next to the skin, silk is cool, breathable and absorbent. But there's no escaping the

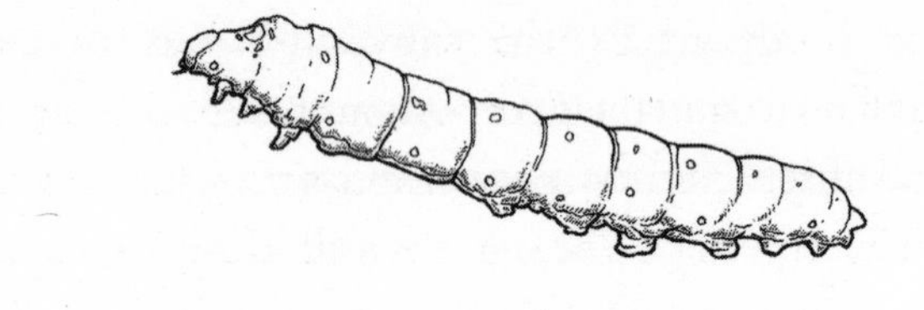

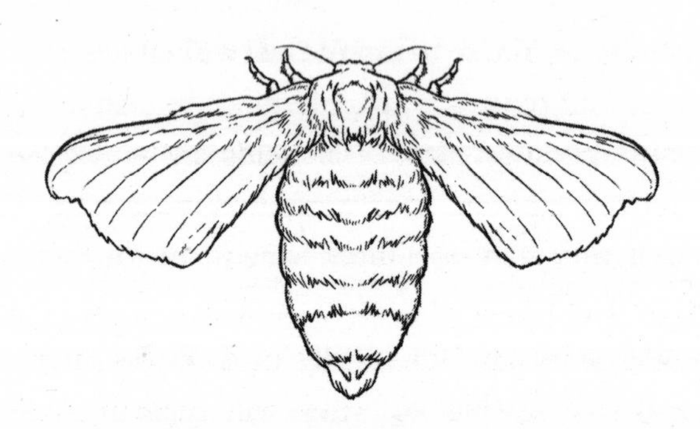

Raw silk is produced by the salivary glands of silk moth caterpillars. The domestic silk moth has been reared in captivity for so long that the adults rarely fly.

fact that what you're wearing is little more than insect spittle.

Besides silk, we have also used insects to obtain dyes. The bright-red dye known as carmine has been used since the second century and comes from carminic acid, a bitter-tasting defensive compound made by the cochineal insects that live on certain species of Mexican and South American cacti known as prickly pears. The bugs are harvested by brushing them off the cacti and drying them in the sun. Carmine has been superseded by modern dyes in the food, cosmetics and pharmaceutical industries, but the natural dye is still available.

✦ Science's little helpers ✦

Most of our understanding of fields such as genetics, physiology, behaviour, biogeography and ecology have come from studying insects, but one insect in particular has been of immense service to us. Many of our homes will have been visited by hordes of tiny flies that seem to appear out of thin air. They will typically be seen flitting around a fruit bowl. How did they get in and where did they come from? A little investigation will reveal an item of fruit at the bottom of the fruit bowl that is definitely past its prime. Try to catch one of the tiny flies in a little tube and look at it closely. It is likely to be a species called *Drosophila melanogaster*, the most important fly in the world. Popularly known as fruit flies, *Drosophila* species have an illustrious history as the pre-eminent model organism in the whole of biology. Early work on genetics used mice, rats, rabbits, guinea pigs, pigeons and even horses, but these animals took up a lot of space, were expensive to maintain and involved long generation times that slowed the process of getting results. Mice, for instance, produce only about six generations in a year; while *Drosophila* can produce no less than twenty-six generations, meaning that researchers can make rapid advances.

Huge numbers of *Drosophila* can be reared very easily in glass bottles, where the larvae are fed on a porridge-like growth medium. There are lots of other reasons why these flies have proved to be such a useful laboratory animal. Males and females are easy to tell apart, and they have only four pairs of chromosome – making life very much easier for geneticists, as mutations can be produced with ease.

You might well ask why fruit-fly genetics are useful to us. Well, 60 per cent of the genes found in the flies are also found in humans. Not only that, but three-

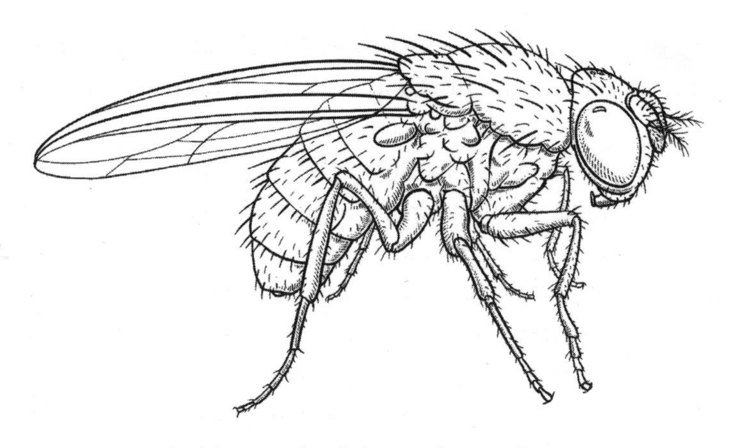

The laboratory fruit fly has revolutionized our understanding of genetics.

quarters of the disease-producing genes in humans can also be found in fruit flies. Through the fruit fly we are beginning to understand the complexities of human conditions such as diabetes, cancer, Parkinson's, Huntington's and Alzheimer's disease as well as how and why we age.

We still have a lot to learn from insects; and studying how they live and survive can give us new ideas to help solve technical problems. For example, some wasps lay their eggs inside the larvae of other insects that themselves feed burrowed deep inside wood. To get to their target, the wasps not only have to drill through the wood but also locate exactly where the larvae are feeding. The way the wasp's egg-laying device (ovipositor) works has proved useful for medical engineers working on steerable probes to deliver drugs to precise locations in the body, or to remove cancer cells; so it probably won't be long before something similar is in routine use. Insects have been around for a long time, so it's no surprise that evolution has come up with lots of solutions for life's little problems. What animals, besides humans, build air-conditioned homes, use underwater breathing apparatus and chemical weapons, make acoustic amplifiers and paper, take slaves, cultivate gardens and farm other animals?

• Bad bugs •

There are, of course, some bad bugs around and it would be remiss not to acknowledge that not all insects are beneficial to us humans. Blood-feeding insects – or rather, the pathogenic micro-organisms they unwittingly transmit – are responsible for innumerable plant, animal and human diseases. About one in six human beings alive today is affected by a disease transmitted by insects. Plague, sleeping sickness, river blindness,

Chagas' disease, yellow fever, epidemic typhus, loiasis, filariasis and leishmaniasis all continue to bring untold suffering and death to millions of people worldwide. Malaria, the most well-known insect-borne malady, is an ancient disease and mentioned in texts as long ago as 2500 BCE. It is caused by protozoan blood parasites in the genus *Plasmodium*, which are transmitted through the bites of *Anopheles* mosquitoes. Malaria infects as many as 500 million people today, but we are at long last making advances in the fight against it. Two vaccines are now available and being used in areas of Africa where malaria is endemic.

Some insects have changed the course of human history. Everyone will have heard of the Black Death, an outbreak of plague, which is thought to have killed one-third of the human population of Europe between 1346 and 1353. Plague is caused by a bacterium called *Yersinia pestis*, which is carried in the gut of a flea. After the flea has fed on an infected rodent and then bites a human, it regurgitates a small amount of infected blood. Plague victims suffered fevers, vomiting, seizures, delirium, coma and eventually organ failure and death. Body extremities appeared black as the skin tissue died. The Black Death was definitely not a pleasant way to go, but it did bring about a series of religious, social and economic upheavals that had far-reaching and generally positive effects, so you might say it wasn't all bad.

Head lice are just plain annoying. These wingless parasites, no more than 3 millimetres (0.1 inches) long, live their entire life on our scalp, holding onto our hair with specially adapted claws and feeding on our blood. The female lays three or four eggs a day near the base of hairs, sticking them on with a strong glue. When the nymphs hatch, they feed and go through three moults before becoming adult. They are not known to transmit disease and are therefore very different from the almost identical body louse, which lives on clothing and most certainly does transmit disease – epidemic typhus being the most notable. This disease, caused by the bacterium *Rickettsia prowazekii*, has caused serious epidemics, especially during times of war when insanitary and overcrowded conditions aided the spread of body lice. With a mortality rate of up to 60 per cent, epidemic typhus was very much more common before the advent of modern antibiotics, but it hasn't gone away.

Bed bugs seem to have made a comeback recently. The Common Bed Bug originally fed on the blood of roosting bats or birds that built nests in caves, and probably switched to human hosts when they settled in caves with fires for warmth during cold spells. I had

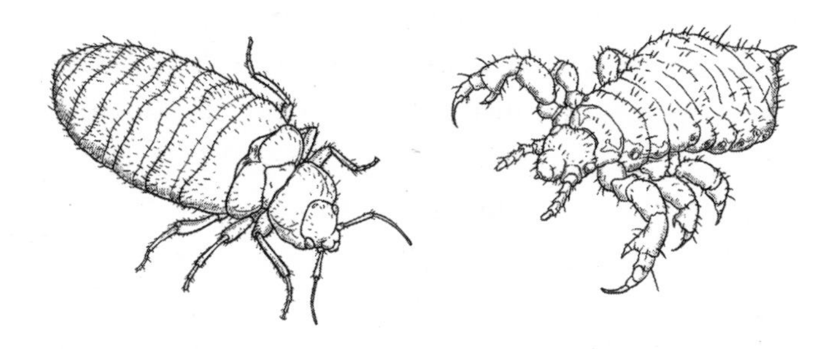

The bed bug (left) and the head louse (right) have a long association with human beings.

the pleasure of going to bed with several hundred bed bugs as part of a television programme about insects in our homes. As I watched the apple-pip-sized bugs being gently shaken onto the duvet's surface, it was clear that they knew exactly where I was: the bloodthirsty horde started to advance towards me almost immediately. I did manage to get some sleep that night but the next day was surprised to find not a single red mark on my skin. It was impossible to believe I had escaped their attention. I had not escaped, of course – I simply had a delayed reaction to their bites and my skin was peppered with itchy red spots the following day.

I worry about the decline of biodiversity and the effects of climate change that are being felt by millions of people already. These are big issues that we need to address very soon or the world as we know it will become largely uninhabitable. I do take a little comfort in the fact that while we may not do very well as a species in the coming years, many insects will survive and continue to evolve – after all, that's what they're very good at.

CHAPTER 3

Insects Under Threat

'Without a biosphere in a good shape, there is no life on the planet. It's very simple. That's all you need to know.'

VACLAV SMIL

Insects are having a very tough time of it in today's intensively farmed and over-developed world. Loss of wilderness and the effects of pesticides associated with industrial-scale agriculture are taking a heavy toll on their populations. Did we really imagine that we could fell the forests, drain the wetlands, burn all the coal and oil and dig up all the peat without there being any negative consequences?

In the past fifty years there has been a marked decline in the abundance of insects all over the world. These changes probably began much earlier and were gradual. It would have been glaringly obvious if insect numbers had plummeted in the space of a year, but a steady, slow decline is not that easy to spot. This is a phenomenon called the shifting baseline syndrome, which basically means that the small changes that happen each year become the new normal. It is only when the cumulative changes become large that people sit up and take notice.

Many scientific studies have highlighted the decline of insects, reporting enormous drops in the abundance of insects occurring in the space of only a couple of decades. This has become headline news; and while it is good to see insects taking centre stage, of course it wasn't long before some people began to poke holes in some of the more dramatic claims. There were questions about exactly how the studies were carried out and how the results were analysed, but at least a vigorous discussion was taking place.

In the vast majority of studies from around the world – whether looking at the abundance of one group of insects, such as dragonflies, beetles or butterflies, or comparing the total amount of insect biomass in the world – the trends have been downwards and we should be paying this much more attention. In North America, for example, iconic species such as the Monarch Butterfly have seen dramatic declines in recent years. While there has been a lack of general large-scale biodiversity surveys, specific insect groups such as tiger beetles and fireflies have declined, probably due to the rise of off-road vehicles and light pollution respectively. Most of the evidence for insect declines have so far come from the UK and Europe, but as more surveys are carried out it seems that the phenomenon is widespread. An interesting study carried out in the deciduous forests at Hubbard Brook in the American state of New Hampshire showed that between the 1970s and the 2010s, there had been a more than 80 per cent decline in the number of flying beetles that were trapped on site. A 2020 study carried out across many sites in the US, however, showed decreases in abundance and diversity among some groups of insects at some locations while others showed an increase or were stable. Taken together, it was concluded that there was no overall decline and that insect populations in the US seemed to be pretty robust.

The detailed picture can be confusing, so some scientists call for more studies to be done and more data to be analysed while others insist that we already know enough to be doing something about insect decline. What does seem clear, however, is that we should at very least be aware of the negative effects of what we are doing right now, and make changes to try to minimize them.

We know that insect populations are greatly affected by weather variations and their numbers often fluctuate from year to year, so it can be difficult to understand exactly what's happening without long-term baseline data for comparison. If only countries had carried out standardized national insect surveys over many years, we would have been able to follow any changes in insect abundance much more easily and accurately. Perhaps we never imagined that we would have to, or perhaps we just don't appreciate what the consequences of insect decline will be.

✦ Where have all the ✦ insects gone?

Aside from these studies, there is very strong anecdotal evidence of declining numbers and disappearing insects. Many people will remember how many years ago the fronts of cars would be heavily splattered with the remains of insects after a summer drive. As a boy I often took on the job of cleaning the family car and can remember quite clearly the large number of insect remains stuck to the headlamps and in the radiator grille. Likewise, driving home at dusk through clouds of moths seems fanciful now but it used to happen frequently.

Another anecdotal example of insect decline is the number of butterflies attracted to *Buddleja*, popularly known as the butterfly bush. In late summer, their blooms attract butterflies, bees and hover flies to feed at their flowers. One record-breaking sighting was of fifty butterflies of ten different species feeding at a single bush. Observations like this may well be a thing of the past. Speak to anyone who grew up in the countryside around the middle of the last century and you will hear stories of the incessant hum and buzz of bees and the chirruping of grasshoppers and crickets. Silent spring in many parts of the world has been replaced by silent summer.

So, what has changed? The biggest single threat to insects – and all other terrestrial species, for that matter – is the loss, fragmentation and degradation of natural habitat. Our planet is 71 per cent saltwater and only 29 per cent land. Some of this land is covered by glaciers, bare rock, salt flats and deserts and is to a

The numbers of butterflies feeding at *Buddleja* has seen a decline in recent years.

large extent inhospitable and uninhabitable, although some specialized life forms do exist even in these most unpromising of places. Before the advent of agriculture around ten thousand years ago, more than half of the Earth's land was forested and the rest was wild grassland and shrubland. But forests have shrunk by one-third and wild grassland and shrubland have lost three-quarters of their original area. What has increased dramatically is cropland, land for grazing livestock and built-up land; between them these now represent nearly half of all the habitable land on Earth. Almost unbelievably, the biomass of human beings and our livestock now makes up 96 per cent of the total mammal biomass on the planet – wild mammals make up just 4 per cent. The success of our species has come at a colossal cost to wilderness and biodiversity.

When you consider what has been happening to the most complex ecosystems on Earth – the tropical rainforests of South America, Central Africa and Southeast Asia – the sheer scale of the destruction is quite simply staggering. Despite growing calls to protect biodiversity, the last twenty-five years have seen the loss of one-tenth of global wilderness areas, with the biodiverse tropical moist forests bearing the brunt of the destruction. And it's not just the rainforests

– by 1950, around 65 per cent of Mediterranean and temperate forest had been lost. These forests of course had a unique insect fauna, which has been seriously impacted. The British Isles has already lost most of its temperate rainforest; but tiny pockets still survive and, with a bit of care, might be able to spread once more.

It seems strange that after all our scientific and technological achievements to date, we do not know how many species live on Earth. The answer will never be known because the decline of global biodiversity has been taking place at such an eye-watering rate. Indeed, the loss is so enormous that many scientists have described what has been happening as the Sixth Mass Extinction Event. The previous mass extinction events were entirely due to natural geological processes, but this one is down to us. As a result of timber extraction, mining, the conversion of forests for agriculture and cattle ranching, the once extensive tropical forests that formerly covered as much as 14 per cent of the land's surface area now cover less than 6 per cent. Madagascar has already lost most of its original rainforest cover and at the current rate of deforestation Borneo, Malaysia, Papua New Guinea and Indonesia will lose most of theirs before the end of this century. If the rainforests go completely, we will lose half or more of all life forms on Earth. We are draining the life blood of terrestrial ecosystems. Now, that's what I call a mass extinction event.

✦ A habitat for all seasons ✦

Habitat loss is going on right under our noses. Forests are lost one tree at a time, and by ignoring small-scale losses we normalize and facilitate a spiral of increasingly large losses.

There are few better opportunities for insect spotting than wandering slowly beside a hedgerow. Of course, hedgerows vary enormously in their value for insects and wildlife, but generally a good hedge is tall and broad at the base with a wide strip of diverse vegetation on either side. There are fruits and flowers in abundance and it is literally buzzing with life – and likely at least one hundred years old.

Hedgerows are historically and culturally significant elements of the landscape in the UK and certain parts of Europe. They have been used to demarcate ownership as well as protecting crops from wild or domesticated animals for thousands of years. Early fields that were hand-cultivated would have been small, but the march of agricultural intensification has seen the size of fields grow to take advantage of the economies of scale and to allow the use of increasingly large farm machinery. These arable fields have no space for nature and have often seen the removal of hedgerows to wring as much profit as possible from the land, resulting in a devastating effect on wildlife. Since the end of the Second World War, the UK has lost half of its hedges. I remember learning about this at school in the 1960s and thinking it would soon be put right. Yet despite numerous grants and schemes to address the issue, the mismanagement and destruction continues.

Hedgerows are an essential year-round resource for all manner of wildlife and act as living corridors that could, if looked after, join up isolated fragments of what natural woodland we have left. Hedgerows support over three-quarters of our woodland birds, half of our native mammals, a third of our butterflies and thousands of other insect species.

A healthy, growing hedge needs to be maintained. Previously, when farm hedges were trimmed with mechanized shears – larger versions of the ones you might use on your garden privet – hedges suffered little lasting damage and the cuttings were allowed to fall to the ground below, where they provided additional structure and microhabitats for wildlife. The invention and widespread adoption of powerful tractor-mounted hedge flails, fast rotating bars bristling with chains or metal bars, has caused monumental damage to hedgerows and the wildlife they support. Flails are used because they are relatively cheap to manufacture and do not need regular sharpening. It has been argued that mechanized hedge flails are not themselves the problem, in the right hands; rather, hedges are flailed far too often, too hard and, despite the regulations that exist, often at the wrong times of year, when birds are nesting or there is a profusion of fruits that could feed a wealth of wildlife. These powerful machines even mangle quite substantial tree saplings, which if left alone could grow clear of the hedge and provide more habitat for wildlife as well as shade. If more care was taken and the regulations were followed, the state of our hedgerows might improve. After all, trees and hedgerows have been shown to be beneficial for farming and the wider environment. They can increase productivity by harbouring the natural enemies of crop-damaging insects as well as the vital pollinators needed.

The Hawthorn Sawfly, also known as the Yellow-Legged Hairy-Clubhorn, is a chunky-looking insect. Its pale-green larvae feed on hawthorn foliage but also on

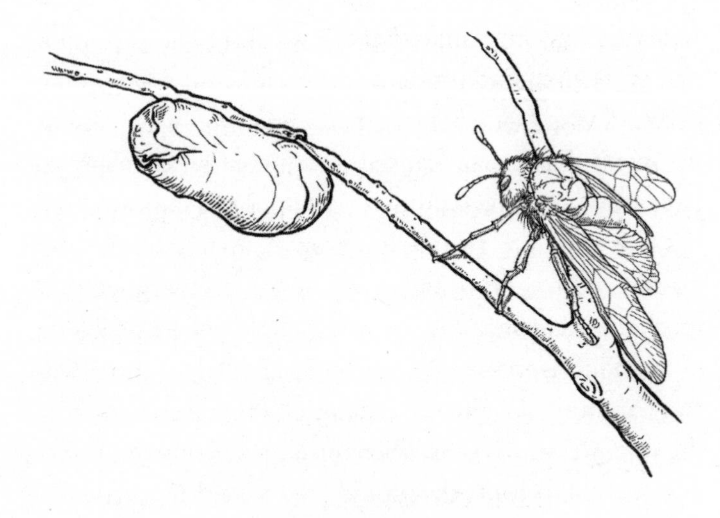

The cocoon of the Hawthorn Sawfly is very vulnerable to hedge flailing.

the leaves of some willows. When fully fed, they pupate in an oval brown cocoon stuck to the ends of twigs. Back in the mid-1800s, many books on nature stated that these cocoons were a common sight on hawthorn twigs in winter. This fact was cheerily repeated in many popular insect field guides right up until the mid-1970s, but no one had thought to check. When I joined the staff of the Oxford University Museum of Natural History in 1984, I wondered why I never saw any of these cocoons on hawthorn twigs. I enlisted the help of some primary schools in the south and west of England and eventually a solitary, empty cocoon was sent to me. The cocoons were not such a common sight after all. I've still never seen one on a twig for myself, and nor have some very committed and energetic entomologist friends of mine. This might have a lot to do with hedge flailing. When neatly sheared twigs fall to the base of the hedge, any insect eggs or pupae present will survive the winter months intact; but flailed hedges are totally pulverized, reduced to so much vegetable pulp that nothing is going to survive. The fate of the Hawthorn Sawfly is just one example and there will be very many others.

✦ For the love of rot ✦

I have never met a child who doesn't enjoy rolling a log over to see what they can find underneath. I still can't resist the temptation to look at the fascinating world of decay. Decay is a fundamental biological process, just as important as life itself. Over 150 billion tonnes (165 billion US tons) of dead wood is produced by forests worldwide every year, but an extensive community of decomposers prevent it from continuing to pile up. It used to be thought that dead and decaying wood was

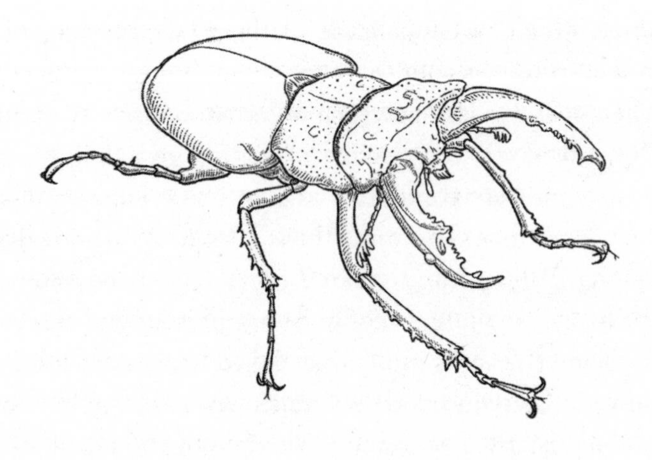

The enlarged jaws of male stag beetles are used in combat to gain access to females.

untidy and not much use for anything, and that it needed to be cleared or burned in case it infected other trees. But the slow decay of wood is essential in returning nutrients to the soil.

The main agents of decay are fungi, but there is also a huge army of saproxylic (rotten-wood-loving) insects whose larvae feed only in decaying wood. About one-third of the rarest insect species in Europe develop in dead wood; and nearly 140 of the 700 European beetles that do are at risk of extinction thanks to the disappearance of veteran and ancient trees. Saproxylic beetles aren't just important in the vital processes of decomposition and nutrient recycling; they are a crucial food source for birds and mammals and some are even pollinators.

The European Stag Beetle, Britain's biggest and most iconic insect species, is confined to parts of the Thames Valley and the New Forest in Hampshire. It is threatened throughout its range by the loss of woodland habitat and the removal of tree stumps, logs and even posts where its larvae develop. The larvae pupate after three or more years feeding underground, but the newly emerged adult beetles face a whole range of threats due to increasing urbanization, including the decking and paving of gardens and the large numbers of cars and domestic cats that kill them. With very similar appearance and ecology, the Giant or Elk Stag Beetle that lives in the deciduous forests of Eastern North America is also suffering declines, and is now recognized as needing greater conservation efforts.

• The bare necessities •

Habitat loss is unequivocally damaging to insect populations. Like all other animals, insects need certain things to survive. They need food plants or a supply of prey and places to lay eggs, which in the case of parasitoids means the presence of specific host insects. Many will be adapted to a particular range of environmental conditions – monthly temperatures, rainfall patterns and so on. If any of their requirements are unmet, the species in question will suffer.

Imagine a generalist beetle that lives among leaf litter or flood debris by the side of a river. It is nocturnal and eats dead or decaying animal matter as well as whatever small creatures it can overpower. It is not very fussy in its requirements and will probably do quite well even if there are some changes to the environment. Now consider a specialist insect. It is self-evident that this insect will fare less well in the face of environmental or other changes. So, what's the advantage in being so fussy? If every insect was a generalist there would be a great deal of competition between species. By specializing in some way, they avoid this competition; but, of course, if they specialize too much, they become vulnerable.

The Purbeck Mason Wasp is one of the rarest insects in the UK.

One such specialist is the Purbeck Mason Wasp that lives in the lowland heaths of Dorset in the UK. Although this wasp can be found in the Mediterranean, Central Asia and North America, it is quite rare in Western Europe and in the UK it is found at only a few sites. The female wasp digs shallow burrows in sandy soil and stocks them with caterpillars of the Heath Button Moth that lives on Bell Heather. Having seized, paralysed and buried as many as twenty of these caterpillars in an underground chamber, she will lay an egg and then seal that chamber before starting a new one. She will continue this process day after day if there

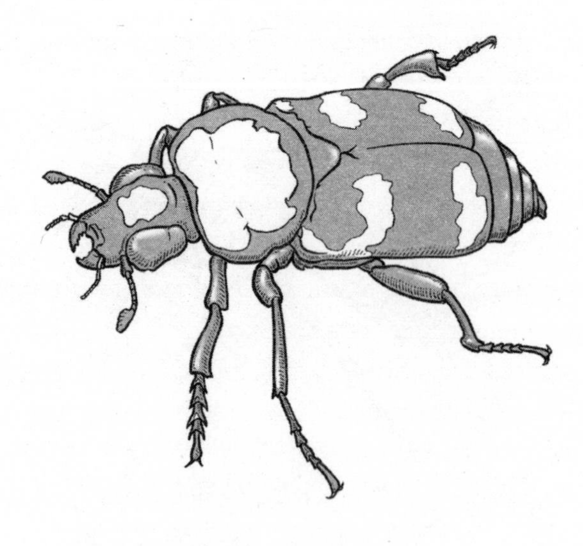

The American Burying Beetle is the largest burying beetle in North America.

are caterpillars to be had as food for her larvae and providing the weather stays fine. But she needs sandy soil, dry, warm weather and lots of Bell Heather for her caterpillar supply, as well as access to water to mix with silt to seal the burrow entrances.

Rarity and eventual extinction can even befall even more generalist species. Take the large (up to 47 millimetres/1.8 inches) and very handsome black, red and orange American Burying Beetle (*Nicrophorus americanus*), for instance. It was once found in thirty-five states across the US, but its numbers have fallen year on year to the point that it can now be found in only a handful of sites in very few states. And unless you really know what you're doing, your chances of ever seeing one alive are very low indeed. Like all burying beetles, this species buries the carcasses of small mammals and birds underground as a food source for its young. The reasons for its dwindling fortunes are thought to be habitat loss, insecticide use and the reduction in the number of animal carcasses.

✦ No more flowers ✦

Fertilizing grassland to feed livestock, especially sheep, has resulted in large expanses of habitat that have little wildlife value. The UK has lost all but 3 per cent of its flower-rich grassland in the last fifty years and the meadows that supported hundreds of species of wildflowers and a diverse insect fauna are now pretty rare. Ten years ago, I was able to film in an ancient meadow that had never been fertilized or sprayed. The variety of wildflowers and the profusion of insects was quite overwhelming. This kind of habitat that we have lost would have been common around the time I was born. The so-called improvement of grassland has played a large part in the decline of insects and some species have been hit very hard indeed. The health and extent of traditional meadows across Europe has been steadily decreasing for a century or more, and some parts of Europe have lost well over 75 per cent of this flower- and insect-rich habitat. The decline is expected to continue due to societal changes, new farming practices and ways of managing the land and the increasing use of chemical fertilizers.

Problems like this have affected whole continents. From the beginning of the sixteenth century, the European colonization of North America brought novel diseases that killed tens of millions of the indigenous population; it also spawned the destruction of the vast prairies through increasingly large-scale agriculture. The Age of Discovery was in reality the start of an age of destruction. The extensive grasslands were ploughed up to grow wheat, corn and soybeans. Natural landscapes teeming with birds and insects, and home to hundreds of wild plant species adapted to their environment, were steadily replaced with crops to feed the burgeoning numbers of settlers. Almost all the tall grass prairie once found in central North America has now been lost: less than 3 per cent remains. And the loss of natural grassland in North America continues, now exacerbated by the effects of urbanization. There are many restoration efforts underway, but these could take decades or even centuries to come to fruition.

It is hard to imagine a summer without the sonorous buzz of bumble bees, but it has certainly got a lot quieter in recent decades. Bumble bees are major pollinators of wildflowers and a very long list of crop species, including broad beans and blueberries, cucumbers and cranberries, pears, peaches, plums and

Once widespread in the UK, the Great Yellow Bumble Bee is now found only in the far north and west of Scotland where suitable habitat survives.

pumpkins. In the 1980s, commercial production of bumble bee colonies for glasshouse pollination got going. Bumble bees can perform the special type of pollination required by tomatoes, which release their pollen only when they are vibrated at about 400 Hz. Bumble bees do this by grabbing hold of the anthers and vibrating their flight muscles without moving their wings (the same technique they use to warm up on cold mornings). Within just three years of commercial rearing being introduced, the Buff-tailed Bumble Bee was pollinating most of the tomatoes grown in Holland. Every time you eat a tomato, you can think of the bumble bees that made it possible.

The UK is home to 10 per cent of the world's bumble bee fauna but sadly they are not being looked after very well: two species are already extinct in Britain and half a dozen more are in serious decline. One of these species, the Great Yellow Bumble Bee, was once fairly widespread across the UK but its range has shrunk alarmingly. This magnificent bee is now found only in the north and west of Scotland, where suitable habitat still survives. I had to travel to Balranald on North Uist in the Outer Hebrides to see one – a newly emerged queen feeding on the flower-rich coastal machair. The fact that few of us today have seen a flower-rich meadow is in itself a problem. You might think that the 'meadows' you see in the countryside are natural, but sadly that isn't the case. They are often heavily fertilized and depleted of nature.

⬩ Poisoning the planet ⬩

Intensive agriculture doesn't just devour wild places; it poisons them and erodes the soil. The use of substances to kill off the insects we don't want to have around goes back thousands of years. Ground-up sulphur dust was used to deter insects by Sumerian farmers more than 4,000 years ago and this was followed by the use of other toxic materials, such as compounds of mercury and arsenic. The insecticidal effects of tobacco and Chrysanthemum plants led to the development of more 'natural' insecticides, but a revolution took place when it was discovered in 1939 that an organochloride compound called DDT (Dichlorodiphenyltrichloroethane) was a very effective insecticide. DDT was widely thought to herald the end of world hunger and disease and was soon being sprayed and dusted all over the place. Advertisements showing happy cows, happy chickens, happy vegetables and happy housewives carried the tagline 'DDT is good for me!'

It wasn't too long before problems arose. Some insects developed resistance to DDT and its damaging environmental effects were laid bare in 1962 by Rachel Carson in her book *Silent Spring*. DDT was an extremely problematic and long-lasting chemical pollutant that was accumulated in the tissues of organisms as it passed up the food chain. It persists in our bodies and soils to this day. Other insecticides were required; and it wasn't long before organophosphates and carbamates came along, followed by synthetic pyrethroids. All these chemicals that interfere with the transmission of nerve impulses are nerve agents (Sarin, VX and Novichok are organophosphate compounds). The trouble is that resistance levels rise as usage is increased – and in any case, as with all insecticides, they kill non-target insects as well. A very real concern is that these chemicals are extremely dangerous to mammals, birds and fish.

Neonicotinoids, a class of insecticides that are chemically very similar to nicotine, were synthesized in the 1980s. These insecticides bind irreversibly to nerve-cell receptors in an insect's nervous system, causing paralysis and death. Their selling point was that they were a lot less toxic to mammals and birds than earlier insecticides had been. These new synthetic insecticides had some other advantages, too. As opposed to contact insecticides, which require reapplication, they were systemic and could be applied as a seed dressing. When the plants grew, the neonicotinoids were taken up, making every part of the plants – including the pollen and nectar – toxic. They were largely used prophylactically – just in case a pest insect should happen to come along. Neonicotinoids soon became the most widely used insecticides in the world, being heavily pushed by manufacturers and touted as beneficial to the environment. In fact, these new systemic insecticides are thousands of times more toxic than DDT.

Alarm bells began ringing only a few years after the introduction of neonicotinoids. In many parts of the world, bees were dying in their millions and it soon became apparent that these insecticides were acutely toxic to bees. As if this wasn't bad enough, researchers proved that even sublethal doses of neonicotinoids could result in serious harm. Bees were suffering from memory loss and severe impairment of their learning and navigational skills, leading to reduced breeding success and increased susceptibility to diseases. Worse still, we now know that on average only 5 per cent of neonicotinoid insecticides are taken up by the crops being treated – the rest is free to contaminate and kill insects in soil and freshwater alike. Freshwater habitats near agricultural fields – even those with protected status – are at great risk from pesticide, fertilizer and livestock waste run-off and communities of aquatic insects are seriously impacted, along with the fish and birds further up the food chain.

Like it or not, there are insecticide residues in everything we eat. In 2017 a survey of honey samples from around the world showed that 75 per cent of them contain at least one neonicotinoid, 45 per cent contained two or more, and 10 per cent contained four or five. The most commonly used neonicotinoids – imidacloprid, clothianidin and thiamethoxam – have now been banned for use on arable crops in Europe because of the harm they do to vital pollinators such as bees, but countries can still use them if they are able to show that there are 'special circumstances'. Since January 2023, it has been illegal for EU member states to grant these approvals. Yet neonicotinoids are still used in many parts of the world where environmental-protection regulations are lax or non-existent. It seems incredible that we can ban these chemicals because we know of the harm they do, then happily allow them to be sold to developing countries – only to import and eat the food they produce.

The next generation of insecticides will very likely also be powerful nerve agents and we will have to go through the whole rigmarole once again. Pesticide manufacturers will claim that we need these chemicals to produce enough food to feed our growing population, but this is a myth. We already waste a third of all the food grown worldwide before it gets near anyone's mouth.

The EU's risk-assessment process is the most rigorous in the world, yet the use of approved pesticides in European agriculture kills a whole range of non-target organisms. In the case of mosquitoes, for example, only a few of the 3,600 mosquito species are a problem and mosquitoes as a group form a large part of many freshwater species' diet. Selectively killing mosquitoes, even if it were possible, would be like whipping out one random card from a very large house of cards and hoping it didn't do too much damage to the structure. As the pioneering ecologist and mountaineer John Muir observed, 'When we try to pick out anything by itself, we find it hitched to everything else in the universe.'

Using anti-parasite treatments such as avermectins to treat livestock has an enormously damaging impact on the abundance and richness of insects associated with animal dung. Just like many antibiotics used in farming today, they are used prophylactically rather than to treat existing problems. There is a really obvious difference between dung pats deposited by untreated animals and those deposited by animals that have been doused or injected with avermectin-style drugs. The dung from untreated animals abounds with beetles and flies and will be broken down and recycled into the soil within days. Dung from treated animals will have no insects at all and the eggs of those that land on it will never hatch. The dung may lie around for many months before the wind and rain eventually breaks it down. And don't forget that the insecticides used to keep your pets flea-free can get into the ponds and waterways they swim in. It has been estimated that the amount of insecticide in one flea treatment is enough to kill 60 million bees; even when diluted, it remains likely to have a significant negative impact on aquatic insects and other invertebrates.

Soil is considered among the most biologically diverse habitats on Earth. It is essential for the survival of plants and by extension most terrestrial life. It not only represents a massive store of carbon but also filters and regulates the flow of water and is home to a rich diversity of life, from microfauna in the form of bacteria and single-celled organisms to vertebrates such as moles, meerkats and wombats. The macrofauna of

quickly stripped the islands of their natural vegetation and all the insects that depended on it. The ships also brought rats with them. This all resulted in the severe impoverishment of the islands' unique biodiversity, much of which would have been found nowhere else on the planet.

The Europeans that colonized Australia brought a whole menagerie of species – livestock, animals to shoot for sport and food and other species that simply served as reminders of where they had come from. We know better these days and many countries have set up biosecurity and phytosanitary procedures that seek to minimize the risks of alien species. Without constant surveillance, however, it is more than likely that some will get though. It is thought that well over 35,000 different species of plant and animals have been translocated to areas of the world they don't belong. In most instances, the species in question don't survive in the new environment, but there are plenty of examples of them surviving and becoming a significant threat to native species.

I remember many years ago finding a rather beautiful metallic blue wasp in a Berkshire garden. It was clearly not a British wasp and, after a bit of work, I discovered it had come from the US, where the species is quite widespread. It looked fresh, as if it had emerged as an adult just a few days before I found it. Now, all I had to do was work out exactly how it had arrived in a garden near Ascot. Heathrow Airport was only fifteen miles away, but I thought the chances of the wasp making it from there unscathed were quite low. The wasp was a mud-dauber, a species that make mud nests that they stock with paralysed spiders before laying their eggs. They are known to coax orb web spiders out into the open before seizing and stinging them. They also take black widow spiders so are thought by many to be a great insect to have around. The nest is then sealed and, when fully grown, the larvae pupate inside the mud structure within a flimsy silk cocoon. After spending the winter like this, the new adults emerge the following spring by chewing their way out. So, all I had it do was find the remains of the mud nest, which I thought must be within a mile of where I found the wasp. The nests are made on buildings, walls, bridges and other structures, often tucked out of the way in overhangs and corners. The answer eventually came in the form of an imported antique sports car (a Lagonda, as I recall). When I ran my fingers under the arch of the front passenger-side wheel, I could feel little lumps of mud. Once I'd carefully prised it all off I could see that just one of the mud cells had been exited, while the others contained dead prey and dead wasps. Although it was very unlikely that this wasp would have survived to become a problem, its journey to the UK could have been easily avoided if the car had been thoroughly checked.

The Box Tree Moth, from Southeast Asia, is not endearing itself to owners of ornamental box hedges and may even threaten the iconic habitat of the eponymous Box Hill. But it's not the moths' fault – the

The Box Tree Moth was accidentally introduced to the UK from Southeast Asia in 2007.

blame lies with the horticultural industry, which quite often accidentally introduces alien species. When the environmental conditions are right, and there's a good supply of suitable food and few natural predators to control invasive insects, the result is rather predictable.

A good example of an invasive insect species is the Asian Hornet (*Vespa velutina*). A fertilized queen wasp is believed to have been translocated from China in a consignment of pottery sent to France in 2004–5. The Asian Hornet is actually smaller than the European Hornet, but what makes them a problem is that they like to catch and kill honey bees to chew up and feed to their young. They can be seen flying around beehives, where they pick off worker bees with ease. In only ten years, the Asian Hornet had spread widely in France and from there made its way into much of Europe. It made an appearance in the UK in 2016, when the first nest was located in Gloucestershire and destroyed. Thanks to increased vigilance, especially from beekeepers, their nests have been located and dealt with; but it is only a matter of time until a nest is missed and eradicating them will become increasingly difficult, if not impossible. Bumble bees generally seem able to deal with the foreign invaders – if they are attacked, they fall to the ground and wrestle with the hornet until it gives up.

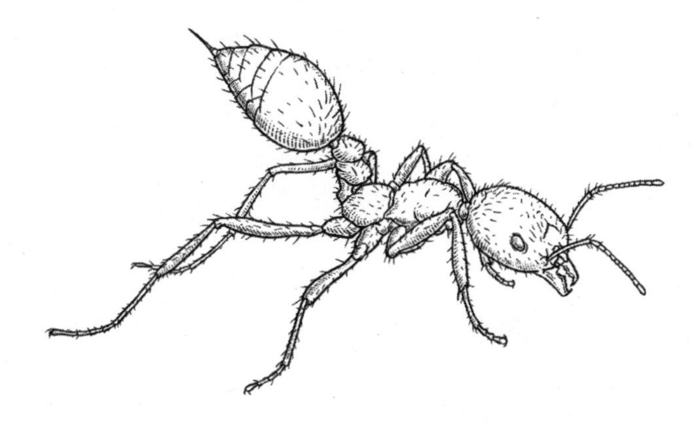

The Red Imported Fire Ant has become a serious problem in many countries where it has been accidentally introduced.

Some insects, such as ants, are very suited to being successful invaders and can get to places without difficulty because they are small and can easily stow themselves away in crates and cargo. In many cases, a new colony can be started by a single fertilized queen. A fire ant from South America, known as the Red Imported Fire Ant, is a classic example of what can happen. Through commerce, this species has been carried to many places outside its natural range, including North America, New Zealand and China. Like many successful invasive species, it is a generalist as well as being able to cope with a range of environmental conditions. In the case of floods, the ants can form living rafts and survive until they reach land. Armed with a bite and a venom-injecting sting, it can quickly overcome most insects and small vertebrates and even drive away ground-nesting birds. Several other ant species have become a serious threat to insect populations in many parts of the world.

New Zealand had no native social wasps until the accidental introduction of two European species. The German Wasp (*Vespula germanica*), which has also spread to North America and North Africa, is thought to have arrived in New Zealand in 1945 in a cargo of aircraft parts. More than likely, a queen wasp crawled into a crate to hibernate only to wake up and find herself on the other side of the world.

The Common Wasp (*Vespula vulgaris*), which has now spread to Australia and South America, was another accidental introduction, having established itself in New Zealand in 1983. New Zealand has a mild climate, an abundance of things for them to eat and nothing that will eat them, so upon their arrival the wasps spread widely and multiplied to become a significant threat to the survival of many native and endemic species. In the beech forests of South Island, higher densities of these wasps have been recorded than anywhere else in the world and it has been calculated that the biomass of the wasps is now greater than that of all the birds and small mammals combined. The wasps eat such huge numbers of insects that their populations are reduced considerably and their survival and that of species higher up the food chain is critically endangered. The whole ecology and natural balance of the forest has been very seriously disrupted and may never fully recover. Biocontrol may offer a solution and the hope is that one or more of the species that naturally control the wasps' numbers in their home range might be used in countries affected by their invasion. Of course, this requires great care to ensure that any biocontrol agent used does what it is supposed to do and doesn't become a problem in its own right.

Invasive plants can also be very damaging to native ecosystems and expensive to control. A very familiar example is *Rhododendron ponticum*, first introduced into Britain in the middle of the eighteenth century as an ornamental shrub and as a rootstock for grafting other rhododendrons. It is fast-growing and has become a very serious problem in much of Europe and the British Isles, where it outcompetes native flora in woodland and heathland and therefore has a very damaging impact on native insect populations. As if that isn't bad enough, rhododendron nectar is toxic to bees. Some races of the Western Honey Bee are tolerant of the toxins in the nectar and will produce honey that can be poisonous to human and lethal to animals.

Another very familiar invasive plant is Himalayan balsam (*Impatiens glandulifera*), which is native to the Himalayas but was introduced into Britain as an attractive addition to gardens in the middle of the nineteenth century. It has since colonized riverbanks and wet wooded areas, where it can cover large areas, displacing native plants and reducing opportunities for native insects. The fact that bees love the nectar may seem like a good thing, but they're so keen on the balsam's nectar that they visit its flowers in preference to native species and their pollination is thus reduced. The plant is now found all over Europe as well as in large parts of North America, which is also now home to several hundred other species of plants that really shouldn't be there and have caused damage to native ecosystems by outcompeting native flora and impacting the local insect fauna. We might bemoan the 'accidental' arrival of unwanted species that wreak havoc on their new home, but in plenty of instances we are directly to blame by introducing species, often for commercial gain, only to find that they become a complete menace.

+ The very hungry ladybird +

We all know that bees are indispensable pollinators, but they can also become a serious problem when they end up in the wrong place. Having proved so successful in pollinating tomatoes and other glasshouse crops in Europe, the Buff-tailed Bumble Bee was transported to many other countries to do the same job. Naturally, it didn't take long before it escaped the confines of the greenhouse and it can now be found in the wild in North Africa, Tasmania and Japan. In South America it has decimated populations of the very large native bumble bee, *Bombus dahlbomii*, by outcompeting it for food and infecting it with a pathogen against which it has no defences.

With every good intention, a species of ladybird with an enormous appetite was picked as a potential biocontrol agent of aphids and other soft-bodied insects that can damage crops. Originally from eastern parts of Asia, the Harlequin Ladybird has been introduced to many parts of the world including North, Central and South America and much of Europe. This 'very hungry ladybird' is just too good at eating aphids and outcompetes native ladybirds – not to mention eating

Not all good news, the introduced Harlequin Ladybird has a voracious appetite for aphids and can outcompete native ladybirds.

their eggs. Worse than that for winegrowers is the presence of harlequin ladybirds in vineyards, because the nature of the defensive secretions they produce has tainted wine. Wherever these ladybirds have become a problem, we now need to think of ways of controlling them without harming native ladybirds. However, the reality is that once invasive species like this get a firm hold it is almost impossible to do anything about it.

✦ Exploitation ✦

You might have seen a certain game show where contestants must undergo trials involving live insects, spiders and other species. By all means, make contestants wallow in liquid manure, rotting food waste or any other sort of foul, stinking gunge you can imagine. Make them face their fears – dark spaces, confined spaces, heights, clowns, whatever their phobias are. But let's not abuse thousands of living creatures for our entertainment. That really stinks.

And then there are the insect collectors, traders in curios and tourist tat such as colourful beetles encased in acrylic resin and tea trays made from butterfly wings. These are things you should never buy, even if the traders claim that the insects have been reared for the purpose. The fact is that many specimens will have been collected from the wild. I'm not talking about scientific collections; rather those held by individuals and collected with little concern beyond owning the complete set of something or having the rarest or most beautiful species. There are global markets for this sort of thing and sadly the protected status of some species seems only to fuel the desire of obsessive collectors even more. As the habitats of rare species are lost, they will become rarer still – and all the more desirable and expensive – until the last one has been caught.

I once took part in a balloon debate where each of the speakers represented an animal or species group. We each had to argue why we should not be thrown out of the basket. In the process, we were acknowledging that, as things stand, we cannot save every species. So, we must decide which species are most important. But on what basis can these decisions be made? Will a species be saved because of their position as a keystone in an ecosystem, without which many more species will be lost? Or perhaps we'll be completely self-interested and keep the ones that benefit humans directly? We should remember that living systems are complicated and we might not fully appreciate the value of the species we think we can do without. As should be clear by now, doing the wrong thing can be disastrous. I represented bees in this debate and was pleased to have won, but the truth is that bacteria or phytoplankton are more important than any of the species we were recommending for the audience's support.

✦ All change ✦

All the threats that insects face will very soon be dwarfed by global climate change, the impacts of which will negatively affect all life on Earth. The nature reserves that have already been set up all over the world may turn out to be useless. We set up these safe spaces to retain and protect the species that became endangered when their habitat had all been cleared or ploughed. The trouble is that the areas are mostly too small and often widely separated. Having a number of small reserves will never be as good as having one really big one.

The dangerous thing about habitat fragmentation is that small areas are much less resilient to change. As the climate warms, what might have been a good reserve for one species will become too hot or too dry and when the species migrate, usually northwards, they will die because the habitat has long gone. Mountains comprise 22 per cent of all terrestrial land surface and will be very much affected by rising temperatures, as will the species that have become adapted to their cool conditions. As conditions warm, they will have to move upwards. However, the area of habitat available to them will become increasingly smaller until it no longer exists.

The biological world has been evolving for a couple of billion years and in that time has produced a rich profusion of species. Most of these species have not survived due to natural processes such as volcanism, climatic instability and asteroid impacts. Even the biggest species ever to have evolved, the dinosaurs, had their day; and so too will we, because now we have entered the 'Anthropocene' – the age of human beings, a species that can destroy entire ecosystems. Through overpopulation, overconsumption and warming the climate, we are well on track to making the planet unlivable – not only for millions of animal and plant species but ultimately also for ourselves. We are like children playing with something very big and complicated that they don't understand at all. Despite our undoubted scientific and technological abilities, our view of the natural world and how it works is simplistic and naive. The American biologist Paul Ehrlich – who, like many others, has warned repeatedly about the consequences of uncontrolled population growth and the depletion of resources – compares the loss of species from the natural world to randomly popping out rivets from the wings of an aeroplane. The removal of one or two and the plane will probably be fine. Perhaps the removal of fifty rivets or so might be survivable; but there will eventually come a point when so many rivets have gone that catastrophic failure and loss of the plane is inevitable. You could imagine insects as the rivets that hold ecosystems together: with each one that becomes extinct, life on the planet becomes less rich and less resilient.

It is worth bearing in mind that we have no greater right to a place on this planet than any other species, be it a fly, a beetle or a wasp. That we think and act as if we were the only thing that mattered has brought us, and the rest of life on Earth, to the very event horizon of existence, after which there will be no going back. It is my view that the continued loss of insects will inevitably lead to a collapse of the life-support systems upon which we depend. Sadly most if not all governments are committed to 'growth' at all costs – a strategy that on a small planet with fixed resources is not only completely doomed to failure, but will also exacerbate the loss of biodiversity to the point that it no longer matters.

CHAPTER 4

Insects and How We Can Help Them

'We differ from our supposedly conservationist forebears only in our greater numbers, more potent technology for inflicting damage, and access to written histories from which we refuse to learn.'

JARED DIAMOND

It is now well established that global biodiversity is in decline and at least 1 million animal, plant and fungal species are at risk of extinction within the next few decades. Climate scientists and biologists have every right to be depressed, and many are. They see what is unfolding in sharp relief and no matter how much or how loudly they speak out, the required response –

which is often impeded by those with vested interests – is very unlikely to match the scale or speed of the actions we need to take.

History tells us that human civilizations do not go on forever. They arise, they blossom, reach a peak of influence and then they fade and disappear. That the human population is now in ecological overshoot can no longer be doubted. This overshoot has been a little while coming and was to a large degree entirely predictable. Many past civilizations have failed because they lived beyond the capacity of the natural world to support them. We have survived by consuming all available resources. Our wastes, some of which will survive for thousands of years, now permeate and pollute every part of the planet.

Just like the boom-and-bust population cycles of fish, birds and mammals that have been studied by ecologists, we find ourselves at a population level well beyond that which the natural world can support. A decline – or 'population adjustment', as some put it – is inevitable. Nothing short of a complete overhaul of how we live, how we engage with the natural world and how we measure success is essential. If we do not do something about it ourselves a rebalancing will take place, but not necessarily on our terms. We need to understand that it is not the planet that needs saving. Earth has been in existence for 4.5 billion years and life on it has survived. It is *us* who needs saving.

I was born in 1954, when there were 2.5 billion humans on Earth and atmospheric CO_2 levels were 313.2 parts per million. I am now seventy. The population has increased to over 8 billion and atmospheric CO_2 levels have reached 416.43 parts per million. In my lifetime, humans have invented machines for seeing the world and beyond. In my lifetime, humans have made incredible scientific, medical and technoloigical advances that would have been unimaginable to my parents or grandparents, yet we seem unable to look after the planet we live on. And we will not survive without the support of the natural world. Indeed, our future livelihoods, health and wellbeing are totally dependent on nature – and to a large extent that means we will not survive without insects. Insects, on the other hand, are natural survivors and have already made it through several mass extinction events. Barring the planet-killing impact of an asteroid of immense proportions or gigantic volcanic eruptions (some of which are thought by many to be well overdue), millennia from now I'm pretty sure there will still be insects around. Our own survival, however, is less certain.

I cannot solve the huge problems of overpopulation, overconsumption and inequality, but I do want to strike a positive note by at least highlighting what we as individuals can do – and more importantly, what we should *not* do if we want to conserve the most important animals that have ever lived on Earth. Collectively we can have a positive impact; and no matter what the outcome, engaging with nature will benefit us all enormously.

As I have already mentioned, the two biggest drivers of species loss worldwide are the destruction and degradation of habitat and the extensive use of pesticides. To a very great degree, both these things are linked directly to the production of food.

• Eating the world •

If there is just one thing you could do to help conserve insects and the habitats they depend on, it is to become vegetarian. There's no getting away from the fact that the scale of our meat consumption is absolutely staggering. Every day, we slaughter nearly 222 million livestock animals (cattle, goats, sheep, pigs, ducks and chickens) – and that figure does not even include fish.

The biomass of our livestock animals is now ten times greater than all the world's mammals and wild birds combined and the amount of land used to produce this number of livestock is humongous. One hundred years ago, only 15 per cent of the Earth's surface was used to grow crops and raise livestock. Today, well over 50 per cent of land has been taken by over by agriculture. It should be of very great concern that areas of high biodiversity value such as the Amazon, Central Africa and Southeast Asia are being particularly badly affected.

The packaged or processed meat you buy in your local supermarket may very well have come from countries such as Brazil, where Amazonian rainforest has been felled and continues to be felled for raising cattle – an estimated 800 million trees in the last six years alone. Yet we have in our grasp the one tool that could bring it under control: what we buy. We do not need to eat as much meat as we do. Not only would eating less or no meat mean that up to three-quarters of the land currently used to raise livestock could be given back to nature, helping to restore biodiversity; it would also substantially reduce the greenhouse gas emissions produced by the industry. The production of 1 kilogram (2.2 pounds) of beef is equivalent to the emission of nearly 100 kilograms (220 pounds) of CO_2. For grains, the figure is about 1.5 kilograms (3.3 pounds), for legumes and fruit it's about 1 kilogram (2.2 pounds) and for many vegetables it's around 0.5 kilograms (1 pound) CO_2 per kilogram produced. Eating less meat is not only the best way to promote the survival of insects, though – it will also reduce your risk of coronary heart disease and stroke, type 2 diabetes and certain cancers, such as colorectal cancer. Bit of a no-brainer really.

Today the seasonal nature of the food year has all but disappeared. We can now eat anything, anytime and we have become so used to doing so it no longer seems out of the ordinary. This seeming miracle is made possible by importing a lot of food. Fruit and vegetables grown in African and South American countries are flown many thousands of miles to reach our shops. Some fruits are even flown halfway around the world to be processed and tinned in other countries where labour costs are low and then shipped back to be sold in the country that grew them. It is an absurd situation that might make sense to an economist but most certainly does not to an ecologist. But we are the consumers and we have a choice. We can really make a difference by paying more attention to what we buy and making informed decisions.

One way of avoiding some of the pesticides that now pervade our food supply is to buy organically produced food. Nutritionally speaking, this food might be the same as conventionally produced food but at least it does not contain pesticide residues. The body tissues of most humans today contain measurable residues of many synthetic chemicals, the cumulative effects of which we know remarkably little about. Organic food is more expensive than pesticide-doused food but that is almost certainly a factor of scale – the more we buy, the cheaper it will become. Grow some of your own food if you can. Home-grown food does seem to taste better and you have the added bonus of reconnecting with the natural world.

If there is one species on the planet I wish had never existed it is the Oil Palm (*Elaeis guineensis*). This plant has resulted in more habitat destruction and species loss – most of which will be insects – than any other in our history. We have already lost untold numbers of insect and plant species and destroyed the home of many critically endangered and vulnerable vertebrate species, including our very close cousin the orangutan. Human use of this stumpy West African palm is thought to be 5,000 years old but the use of palm oil in processed foods has propelled it to become one of the world's most widely produced primary crops.

Originally grown commercially for use as an industrial lubricant, the oil that can be extracted from Oil Palm fruit is versatile and of high quality. It is now used in all sorts of foodstuffs as well as in soaps, shampoos and cosmetics. It might well be that somewhere approaching one half of all the products you see stacked on supermarket shelves contain palm oil. The world has become well and truly hooked on the stuff. Many will point out that, of course, we need vegetable oils and palm oil is a good thing as the yield of oil per area of crop grown is a lot higher than any other vegetable oil, and the trees need less energy input and less pesticides to maintain them. The major problem is that Oil Palm needs a tropical climate to grow and Oil Palm plantations have caused truly massive deforestation in all the tropical countries where they occur. The world is losing untold biological riches in return for cheap processed food, face cream and biofuel that is no greener than the fossil fuels it seeks to replace. Palm oil may be far cheaper to produce than other vegetable oils, but the cost in the long run will be incalculable.

I have stood at the edge of primary rainforest in Sabah, Borneo and watched the advance of Oil Palm plantations. The contrast between the two sides could

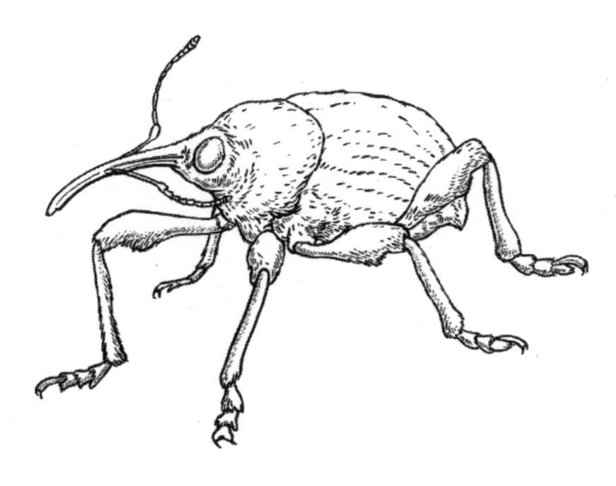

The Red Palm Weevil feeds on a range of commercially important palm trees such as date, coconut and sago.

not have been starker. On one side, a lush, high-canopied forest where insects and other species abound; on the other, a scorchingly hot monoculture of Oil Palm with fewer species than you might find on the margins of a local football pitch. It is still possible to find undescribed species of insects in rainforests, but there will soon come a time when this is not the case.

From an annual production of less than 1 million tonnes (1.1 million US tons) in the early 1960s, palm-oil production has gone through the roof. It is projected that by 2050 the world output will be somewhere in the region of 240 million tonnes (264 million US tons)

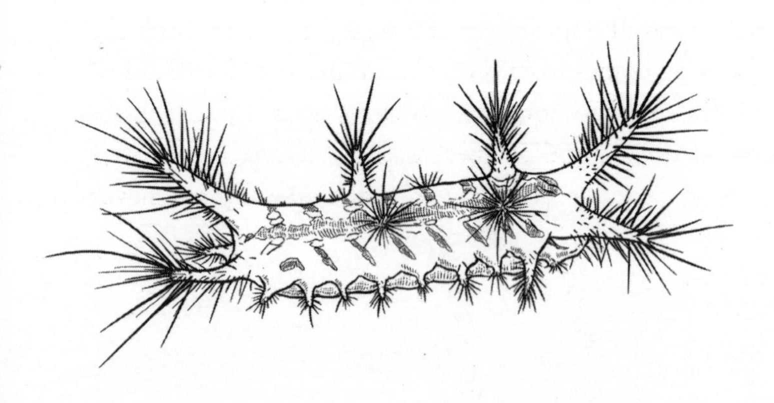

The caterpillar of the Coconut Moth is considered to be one of the major pests of Oil Palm.

a year. As long as the money keeps rolling in, those in a position to exert some regulation will look the other way and the forests will keep on burning.

But you don't have to be a passive consumer. Again, look at the labels, even when some manufacturers don't make it easy. Small print and many confusing pseudo-scientific names obscure the presence of palm oil and its derivatives in a lot of products. What all labels should say is either 'contains palm oil' or 'palm-oil free'. Many people are already very concerned but some are reassured that the product they are buying contains 'sustainable palm oil' and is therefore acceptable. We need to be very clear about this: the sustainability certification for palm oil comes from the palm-oil industry itself; and if Oil Palm plantations are grown on land where tropical forest once stood – whether it was cleared yesterday, last year, or a decade ago – the palm oil is not and never can be described as sustainable.

✦ How does your garden grow? ✦

It might surprise you to know that only 13 per cent of land in North America has any degree of environmental protection. The figure for the world as a whole is close to 15 per cent – and even the level of protection that exists misses out very significant areas of high biodiversity value. Protecting half of the Earth's ecosystems might seem a tall order, but if we do not take action we risk losing a substantial proportion of our living heritage and endangering the long-term health and stability of the natural world.

Do you have a garden, and if so what sort of garden is it? Does it have a striped lawn with neat edges and herbaceous borders with well-ordered clumps of plants? The reason gardens are very important is that they cover a large area, a little over 500,000 hectares

(1,235,527 acres) in the UK alone. To put that figure in context, the UK has around 224 national nature reserves that cover only 90,000 hectares (222,395 acres).

Chief among modern garden crimes against nature is the use of artificial lawn. I find it hard to understand why anyone would want to use this material in a garden. It must be cleaned. It doesn't last that long. Often badly laid, it soon succumbs to the elements and to plants growing in it or through it. It usually ends up in landfill, where it will last for hundreds of years. There was a case of someone asking their local council to cut down a cherry tree that was growing outside their garden because its leaves fell on their artificial lawn and made it look untidy. I'm not going to devote any more words to this topic because I need to keep an eye on my blood pressure these days.

It seems blindingly obvious that to help insects we simply need to stop doing the things that harm them in the first place – and what we do in our gardens is a very good place to start. If you want to encourage insects into your garden, you need to give them good reasons to be there. Those reasons are the presence of suitable plants or prey to eat, shelter and places to lay their eggs.

The first serious study of garden invertebrates was carried out by Jennifer Owen over a period of three decades in a well-stocked garden in Leicestershire in the UK, with over 450 recorded plant species. She racked up an impressive tally of over 2,000 animal species, the vast majority of which were, of course, insects. For the vertebrates, there were sixty-four species, including forty-seven birds and seven mammals, many of these being entirely dependent on the insects present. If you have anything like this situation in your garden, you have all the elements for some very interesting ecology that will be going on right under your nose. I think it's fair to say that there is more going on in a garden than most people would imagine and all it takes to discover the details is careful observation and an inquiring mind. If you are lucky enough to have nesting birds in your garden and a place from which to observe them, you will be astonished by the volume and variety of insects and other small invertebrates the parents skilfully glean from the surrounding vegetation to feed their brood. Many tens of billions of wasps, flies, beetles, spiders and caterpillars are consumed by British nestlings annually.

Jennifer's garden was so rich in species partly because it was completely free from pesticides. Globally, over 3.5 million tonnes (3.9 million US tons) of pesticides are used every year to kill plants, insects and other species regarded as undesirable. Around 450,000 tonnes

(496, 000 US tons) of pesticides are used annually in the US alone; and American homeowners use somewhere in the region of 36,000 tonnes (40,000 US tons) of pesticides. The effects of this chemical onslaught on the natural environment are very significant and almost certainly largely undocumented.

The vast monocultures of global agriculture are one thing but it seems completely preposterous that anybody would use pesticides in their garden when there are dozens of predatory species, ladybirds, lacewing and wasps that will do the job for free. If you have any old chemicals lying around, get rid of them now. Don't just wash them down a drain. These chemicals are dangerous so contact your local authority for advice on their safe disposal.

I must be honest and tell you that my hollyhocks would not win prizes at the local horticulture show, even if they were the only entry. The flowers are rather small and peppered with little brown blotches, the result of weevils feeding on the young flowers. The sight of this would have many gardeners running for their bug guns. But the weevils have given me hours of pleasure. Weevils have a long, slender nose or rostrum, on the end of which are the small mouthparts. My hollyhocks are host to not one but two species: one is a small black

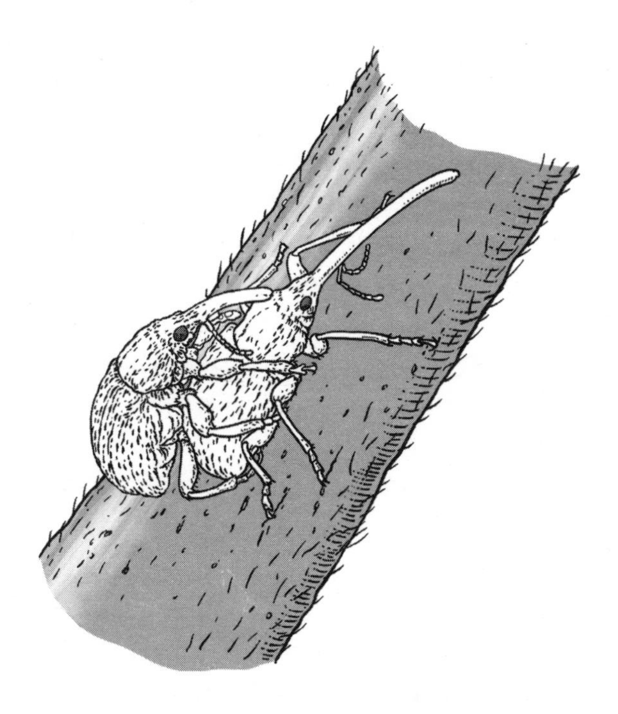

Hollyhock weevils mating. Gardens can be an important refuge for insects in an increasingly agricultural landscape.

weevil also found on common mallow and the other, a relative newcomer to the UK, is larger with orange legs and a remarkably long rostrum. These delightful little beetles trundle about happily on the hollyhocks, mating and feeding on the buds, and I'm glad they're there.

If you're going to have a wild garden, you can

start by not mowing your lawn. In the first year you might only mow paths and a few patches. Remove the turf and fill the patches with the sorts of plants you'd like – or rather, the sorts of plants insects will like. A friend's garden was once a boring suburban patch and is now a haven for insects. It's not something you need to do all at once; instead let it evolve, adding features such as ponds and log piles as and when. A pond will attract dozens of aquatic insect species – water beetles and pond skaters will appear as if by magic as soon as it is filled and later, as the pond matures, damselflies and dragonflies will fill the air above.

What flowers to plant? Certainly, old-fashioned varieties are more attractive to a wide range of pollinators. Plants have evolved flowers with a whole suite of visual and olfactory characteristics as well as rewards of pollen and nectar to attract specific insects and deter others. Some flowers are not too fussy about what comes along while others rely on attracting very particular species. The interactions of flowers and their pollinators can be very nuanced. But in response to the decline in the number of pollinators worldwide, some plants have been shown to reduce the amount of nectar they produce and the size of their flowers. Why waste resources producing something that is no longer

Old-fashioned flowers such as foxgloves are an important food source for bumble bees.

of benefit? Some fear that self-fertilization in plants, if it increases, could further accelerate the decline of pollinators. So, before this ever becomes more than a possibility, let's get the pollinators back.

There is no better sight than bumble bees foraging at foxgloves. Happily, there are plenty of useful websites suggesting suitable flowers for wildlife-friendly gardens, but make sure that there will be something in flower throughout the season so pollinators have a constant supply of pollen and nectar. Planting plug-plants is easier and gives faster results than seeds; and once established the plants will spread naturally into other places. Keep on digging up more patches of the old lawn and planting. You can replace conventional shrubs in the borders with primroses, cowslips and herbs like marjoram, sage or thyme. Some plants can get a bit too successful, so you may need to manage them: pull them out or dig them up and give them away to a good home. Try to maintain a variety of microhabitats such as long grasses or weedy areas for bumble bees to make their nests. Early spring bees need food, so dandelions are very important; and ivy flowers are one of the best sources of nectar towards the end of the year.

Try to do away with tidying your garden too much. Fallen leaves actually form a protective layer for the smallest of insects, and provide a natural shelter for overwintering hover fly larvae, especially those species that hunt aphids in the summer. Thicker layers of leaves are good, so that small species can move up and down in response to changes in moisture and temperature. Leave the cutting back and clearing of seed heads as late as possible (into late February or March except when covering bulbs below) and don't do it all at once.

Hollow, dead stems may be full of hibernating insects and larvae so leave them be for as long as possible.

The only nocturnal illumination you need will be supplied by the moon and stars, so try to minimize the lighting in your garden. I've seen some gardens so brightly lit after dark that you could read without difficulty or even carry out minor surgical procedures – it is wasteful of energy and damaging to wildlife. Recent research has shown that insects, and moths especially, are very negatively impacted by lights at night. Insects are attracted to bright lights and in the tropics I have seen piles of dead moths under street lamps. You might think that would not be an issue in the UK but it is. Recent research has shown that street lighting in the south of England can reduce the abundance of moth caterpillars in grass verges by one-third and in hedgerows by almost a half compared to equivalent unlit roadside habitat. Light pollution, or ALAN (artificial light at night) is on the rise everywhere. One recent study estimated that the night-sky brightness in North America has increased by roughly 10 per cent per year. A glowing sky has been shown to have detrimental effects on species such as bats as well as insects.

Now, let's talk about peat. For quite a while, the horticulture industry said that without peat it would be impossible to grow certain plants, but digging up peat to grow garden plants is just inexcusable. Quite apart from the value of peat bogs as a massive store of carbon, it would be much better to leave them and their unique fauna and flora well alone. I have a simple rule: if you can't find a suitable peat substitute, grow something else.

If you'd like to discover the hidden treasures in your outside spaces a little more, a very good way of recording your findings is to photograph them. Macrophotography need not involve lots of expensive camera gear. Many mobile-phone cameras do pretty well and there are some excellent compact cameras that take phenomenal close-up images. Don't forget the positive power of social media, too. I resisted taking the plunge for many years but soon found that there are an awful lot of people in cyberspace who know a great deal about insects and are generous enough to share their expertise. The more you look for insects, the more you will find – it can become quite addictive.

If space allows, a suitable native tree will give your garden shade. Perhaps an apple or pear tree, and preferably one that grows fruit you'd like to eat yourself. A well-balanced wild garden should look after itself to a large degree, but do remember to have somewhere to sit in the afternoon sun where you can watch and listen to the insects. A very good purchase would be a pair of short-focusing binoculars (they're not just for birdwatching). Your wild garden may take several years to come to fruition but along the way it will give you a lot of pleasure and instruction while it helps the beleaguered insect fauna and boosts local biodiversity. If you don't have a garden, consider joining forces with someone else or helping someone who can no longer manage their plot. Failing this, your local wildlife conservation group will surely welcome you with open arms.

✦ More trees please ✦

If there is one thing we need to do straightaway, it is to plant more trees. Trees are the most remarkable organisms. They are specialists in carbon sequestration, using complex biochemical alchemy to pull carbon dioxide out of thin air as they grow and pump out life-giving oxygen as a waste product. Trees evolved over 350 million years ago and it is highly unlikely that we will ever be able to replicate something even vaguely like a tree. Trees cool the planet, regulate weather patterns and, crucially, are a habitat for countless numbers of insects and other species.

In urban settings the shade trees provide is going to prove vital in the coming years, reducing surface temperatures by as much as 20°C (68°F). In addition to shade, urban trees absorb noise and pollutants, improving human health and wellbeing. But still we cut them down for development, to maximize profits.

The forests of the world hold not only most of the plant biomass on Earth but also most of the species – mainly insects. But today we find ourselves in the middle of a climate crisis and an ecology crisis. That we should be looking after the world's forests is incontrovertible and yet there is much talk of how we might overcome the climate difficulties ahead with technical fixes and hacks that will allow us to continue as we have done, albeit a little more slowly, down the same path towards an unlivable world.

The world has lost between one-third and half of its tree cover since humans spread around the planet and we can ill afford to lose any more. The area of woodland in the UK is around 3.25 million hectares (8,030,925 acres); and compared to the rest of the European Union, which has a tree cover of around 40 per cent, this represents a rather measly 13 per cent of the land area. Forests and woodlands cover over 33 per cent of the United States, while Canada's tree cover is a healthy 50 per cent.

Planting more trees is one of the best solutions to reverse the decline of biodiversity, protect the environment and reduce our net carbon emissions. By this I don't mean slapping in a lot of alien conifers in straight lines. We don't need the wrong trees in the wrong places and we don't need species-poor habitats where species-rich ones once stood. What we need to do is to rebuild our natural woodland stock and with it all the species that live there. When mature trees are felled we are told not to worry – more trees will be planted. But when a mature woodland is lost, so too is the rich community of species that has accumulated over hundreds of years. Regeneration is the best solution, so we need to reduce grazing pressure and damage and let our woodlands and hedgerows regenerate. Some regeneration schemes have already started, though it may be some years before we see their benefits. Regardless, we need thousands more schemes of all sizes. If a good local seed bank still exits, assisted regeneration will cost a lot less than frantic tree-planting and will be more successful in the long run. We've pretty much wrecked the ecology of many parts of the world in the last couple of hundred years; but if nature is one thing it is resilient, and if we re-establish the foundations of natural woodland it will come back.

In my university holidays I would often go to the Scottish Borders, where I would stay and help out on a sheep farm in the Moffat Hills. Not far from where I stayed is the Grey Mare's Tail, a wonderful 60-metre (197-feet) waterfall fed by Loch Skeen, now home to the Vendace, one of Scotland's rarest freshwater fish. Like the rest of the Southern Uplands, the land all around is treeless, boggy moorland and when I went on long hikes it never occurred to me that it had not always been like this. The truth is that the bleak landscape is the result of centuries of constant nibbling by sheep, goats and deer. If we're going to stop and reverse the decline of insects – of biodiversity in general – we need to rebuild damaged and fragmented habitats from the ground up. To my mind, trying to save some endangered species is rather futile if you ignore the rest of the ecological community around it. Such efforts will amount to nothing because, as we should know by now, everything in nature is connected.

Not far from the Grey Mare's Tail is a tremendously inspiring example of ecological restoration – Carrifran Wildwood. This is a story of hope and the passion of a small group of people who were dismayed by the state of the natural habitat and decided to act. Their aim was to restore a once wooded valley that would have flourished before humans settled in the area and started to graze animals around 6,000 years ago. Land covering some 655 hectares (1,619 acres) was purchased by public subscription and all grazing animals were removed. At the turn of the millennium, when I was running about catching insects in the Hunza Valley in Pakistan (another massively overgrazed part of the world), the first of over 750,000 trees were planted in Carrifran Wildwood. In total, about thirty different kinds of trees and shrubs were planted, including downy birch, sessile oak, rowan, aspen, wych elm, hazel, holly, hawthorn, blackthorn and juniper. These were not just any old trees; rather, they were the offspring of local species that still survived in the area, clinging to rocky gullies and other areas inaccessible even to sheep and feral goats. Millions of seeds were collected and grown on, ready to be planted out. Twenty-four years later, the trees are thriving, the insects are returning and the ecology is restored. The project, now expanded by the Borders Forest Trust to a total of 31 kilometres^2 (12 miles^2) of remote hill land shows that with a plan and a platoon of volunteers, anything is possible. I hope this might inspire you to think about what you can do in your area. Little bits of ecological restoration here and there soon add up.

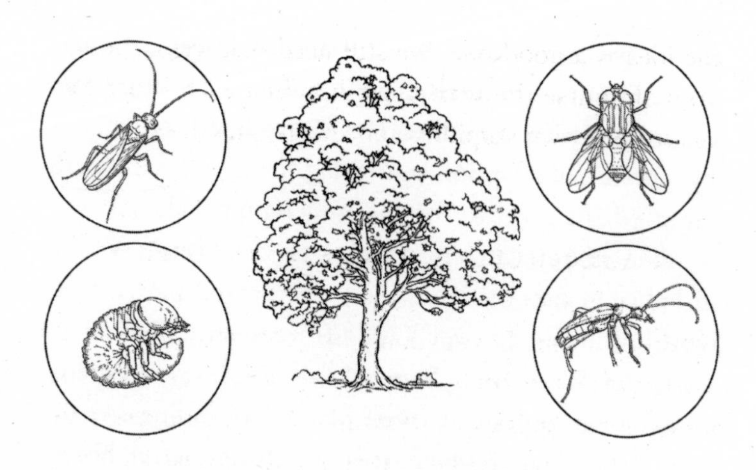

A tree can be host to many hundreds of species of fascinating insects.

Perhaps we all need to be more like Cuthbert Collingwood, Nelson's second-in-command at Trafalgar. When Collingwood was on leave at his home at Morpeth in Northumberland, he loved to walk over the hills with his dog. He would take a pocketful of acorns and whenever he saw a suitable spot for an oak tree to grow, he would push an acorn into the soil. Many of the acorns Collingwood planted grew and some are still alive. Collingwood's purpose was a little niche – to ensure that there would always be a supply of stout oak to build the ships that ensured the country's safety – but the idea is a good one. We still need oak trees, though now of course the war is not between each other but against our own stupidity and short-sightedness.

✦ Get active and speak out ✦

Don't be silent. If you don't like the way your local road verges are being mown too early or too often, for example, speak out. If you keep quiet, no one will know what you think. A very good way to let your voice be heard is to join your local Wildlife Trust or naturalist group. By volunteering your services, you will become part of the wider effort and impetus to restore nature and wild spaces. We don't just need to protect woodlands and let the trees grow old; we also must look after meadows, peatlands and wetlands. We need to take much greater care of the soil, our hedgerows and our waterways. Just think of the advantages if we all give up some of our time to do something positive where we live. We will be healthier and happier, and insects will thrive. There is great strength in numbers and together we can get bigger and better things done. If we do not become activists, we are simply docile recipients of the misguided or malign actions of others.

Humans tend to be reactive, not proactive, but now we face problems so immense that they can no longer be ignored. In the last few years there have been several major reports from various intergovernmental bodies and a growing volume of academic research that all point to the fact that we urgently need to change our ways or the here and now may well become the hereafter.

I used to think that science and common sense would save the day, but I realize now that I was being naive. It's up to every one of us to make a difference where we can.

It is estimated that annual bank loans and subsidies linked in some way to ecosystem and wildlife destruction are in the region of $5–7 trillion per year. Banks around the world lend money to fund all sorts of projects – some good, and some not so good. We may not necessarily connect the ways in which we invest our money with the protection of the natural world, but it is something we do have control over that can collectively make a huge difference. Ethical banks committed to never funding things such as deforestation exist, so go and find one.

Rather than sitting and watching as things get worse, now is the time to change what we do; to re-evaluate what is and isn't important with an understanding that a future without insects will not be much of a future at all.

+ Mushi-yoku +

You might have heard of Shinrin-yoku, the Japanese practice of forest bathing. It's seen a resurgence in interest over the last few years, and there are many books on this subject. For the uninitiated, forest bathing essentially involves spending time in a forest or woodland where you walk slowly or sit quietly and try to become gradually more aware of your surroundings. Much has been written about the physical, mental and emotional benefits of Shinrin-yoku. Practitioners are convinced that simply sitting quietly and experiencing the sights, sounds and smells of this natural habitat can help reduce depression, anxiety, stress and insomnia and lower blood pressure. Even if you've never practised Shinrin-yoku, you're probably familiar with that feeling of wellbeing you get after a woodland walk? It may very well be brought about by certain natural odours such as geosmin (from the Greek meaning 'earth smell'), a ubiquitous organic compound that gives soil its characteristic earthy odour.

Now, I'd like to introduce you to a slightly different way of slowing down and getting in touch with nature that I practise myself. I call it insect bathing – Mushi-yoku (the Japanese word might also translate to 'insect desire'). I am rarely happier than when I take a walk in an ancient woodland or along a wild coastal path in search of insects. Progress can be slow because when I say walk, what I actually mean is a *really slow* walk where I can take time to look closely at every single beetle, bee or bug that I find along the way.

I sometimes take a sweep net – a common entomologist's collecting tool – so that by making my way through the vegetation I can bag a lot of insects in a relatively short time. The downside of this mass-collection approach is that it tells me only that some insect species or other is present and perhaps how common it is at that location. I will have absolutely no idea what anything in my net was doing or how it was interacting with other species – I won't even know what plant they were feeding or sitting on. Much better to find a small colony of ants living inside an acorn cup or some beetles scuttling between the gills of a fungus. By simply looking intently at a cow parsley flower, or a dung pat or what's going on underneath a rotting log, you will begin to understand the secrets of the insect world. All the while, you will probably find yourself asking questions. What are they doing and why? While such a methodical approach might not exactly give you a cardiac workout, I guarantee you will feel calmer and more connected with the natural world at the end of it.

Epilogue

What does the future hold? I only wish I could tell you but one thing I can say for sure is that if we continue to live our lives *apart from nature*, not as *a part of nature*, it's going to be difficult, to say the very least. We have, for a very long time, failed to grasp the simple fact that our continued existence, our health and our wellbeing, depend entirely on the survival of the natural world. Without the rich diversity of life (mostly insects remember) on Earth we will surely decline. The loss of biodiversity has been accelerating in recent decades, prompting some to suggest that what we need to do is figure out exactly which species are essential to our survival and which species are not. I'm sorry to have to tell you that we're really not that smart.

In the last seventy years though, the time I have been alive, we have made immense scientific and technological advances. We rely on electronics to live our lives and a little while from now artificial intelligence will rise to dominance for good or ill. We live longer thanks to biomedical science. We have deciphered and can manipulate DNA, the molecular code of life. This has revolutionized biology immeasurably and has even extended my own life.

In my lifetime, humans have invented new ways of observing the world. At one end of the spectrum giant machines have proved the existence of infinitesimally small subatomic particles, and at the other end we have made earth-bound and space telescopes that have allowed us to see back in time, almost to the very beginnings of the universe.

Fifty-six years ago, when I was a teenager, astronaut William Anders, on board Apollo 8, took a photograph of Earth as it rose into view from behind the moon. *Earthrise*, as the image became known, was a wake-up call for many and proved influential in bringing environmental issues and concerns to a much wider audience. Perhaps that was the time we should have realized that what we saw, a small and rather beautiful planet, was worth looking after. We have had many wake-up calls since but have pressed the 'snooze button' every time. So why, when the evidence of severe damage to the biosphere is now incontrovertible, do we not take decisive and immediate action to try to avert the worst effects of what we alone have brought about? Why are we dragging our feet over the environmental issues that will blight the lives of generations to come? Is it because we really don't believe that the threat exists? Is it that the task is now so big that is seems undoable? Do we imagine that science and technology will win the day somehow? Or are we simply afraid that if we take action and no one else does, we will be worse off?

We need transformative change and better ways of measuring our success. We need to consume less, waste less and we need to reduce our population. We have a huge task ahead of us but we must not let its enormity paralyse us into inaction, for if we do nothing mass extinction and climate upheaval will be inevitable.

Perhaps I should have been an archaeologist, digging in the dirt to find clues to the lives of those who went before us, or an astronomer discovering the secrets of the stars. I could have lost myself in these subjects where the things I was studying were not disappearing before my eyes. As Aldo Leopold, the pioneering American conservationist once wrote, 'One of the penalties of an ecological education is that one lives alone in a world of wounds.'

George McGavin

About the Author

George McGavin studied Zoology at Edinburgh University, followed by a PhD in entomology at Imperial College and the Natural History Museum in London. After twent-five years as an academic at Oxford University he became an award-winning television presenter. George is an Honorary Research Associate of the Oxford University Museum of Natural History and an Honorary Principal Research Fellow at Imperial College. George is also a Fellow of the Linnean Society, an Honorary Fellow of the Royal Society of Biology and an Honorary Life Fellow of the Royal Entomological Society. As well as his many TV documentaries, George has written numerous books on insects and other animals. In 2019, he became the President of the Dorset Wildlife Trust.

Acknowledgements

I am most grateful to Anne Riley of the Wharfedale Naturalists for reading and commenting on drafts of the book; and to Karen King, who has done a marvellous job of the illustrations.

Selected Reading

There are so many good books on insects these days – evidence, I hope, of the increasing level of interest. There are field guides to the insects of many countries and regions, but even the fattest of them cannot hope to be comprehensive in their coverage. I have included a handful of books here if you would like to learn about the biology, classification and importance of insects in more depth.

Ashmole, P. and Ashmole M., *A Journey in Landscape Restoration: Carrifran Wildwood and Beyond* (Whittles Publishing, 2020).

Chapman, R. F., Simpson, S. J. and Douglas, A. E., *The Insects: Structure and Function* (5th edn) (Cambridge University Press, 2013).

Chittka, L., *The Mind of a Bee* (Princeton University Press, 2022).

Goulson, D., *Silent Earth: Averting the Insect Apocalypse* (Vintage, 2021).

Goulson, D., *The Garden Jungle: Or Gardening to Save the Planet* (Jonathan Cape, 2019).

Grimaldi, D. A., *The Complete Insect: Anatomy, Physiology, Evolution, and Ecology* (Princeton University Press, 2023).

Leather, S., *Insects: A Very Short Introduction* (Oxford University Press, 2022).

McAlister, E., *The Secret Life of Flies* (Firefly Books, 2017).

McGavin, G. C. and Davranoglou, L. R., *Essential Entomology* (2nd edn) (Oxford University Press, 2022).

Ollerton, J., *Pollination and Pollinators: Nature and Society* (Pelagic Publishing, 2021).

Owen, J., *Wildlife of a Garden: A Thirty-year Study* (Royal Horticultural Society, 2010).

Sumner, S., *Endless Forms: The Secret World of Wasps* (William Collins, 2022).

Index

Page numbers in *italic* refer to illustrations